T0344485

SAFETY MANAGER'S GUIDE TO OFFICE ERGONOMICS

SAFETY MANAGER'S GUIDE TO OFFICE ERGONOMICS

CRAIG CHASEN

WILEY

A JOHN WILEY & SONS, INC., PUBLICATION

Copyright © 2009 by John Wiley & Sons, Inc. All rights reserved.

Published by John Wiley & sons, Inc., Hoboken, New Jersey.
Published simultaneously in Canada.

No part of this publication may be reproduced, stored in a retrieval system, or transmitted in any form or by any means, electronic, mechanical, photocopying, recording, scanning, or otherwise, except as permitted under Section 107 or 108 of the 1976 United States Copyright Act, without either the prior written permission of the Publisher, or authorization through payment of the appropriate per-copy fee to the Copyright Clearance Center, Inc., 222 Rosewood Drive, Danvers, MA 01923, (987) 750-8400, fax (978) 750-4470, or on the web at www.copyright.com. Requests to the Publisher for permission should be addressed to the Permissions Department, John Wiley & Sons, Inc., 111 River Street, Hoboken, NJ 07030, (201) 748-6011, fax (201) 748-6008, or online at http://www.wiley.com/go/permission.

Limit of Liability/Disclaimer of Warranty: While the publisher and author have used their best efforts in preparing this book, they make no representations or warranties with respect to the accuracy or completeness of the contents of this book and specifically disclaim any implied warranties of merchantability or fitness for a particular purpose. No warranty may be created or extended by sales representatives or written sales materials. The advice and strategies contained herein may not be suitable for your situation. You should consult with a professional where appropriate. Neither the publisher not author shall be liable for any loss of profit or any other commercial damages, including but not limited to special, incidental, consequential, or other damages.

For general information on our other products and services or for technical support, please contact our Customer Care Department within the United States at (800) 762-2974, outside the United States at (317) 572-39983 or fax (371) 572-4002.

Wiley also publishes its books in a variety of electronic formats. Some content that appears in print may not be available in electronic formats. For more information about Wiley products, visit our web site at www.wiley.com.

Library of Congress Cataloging-in-Publication Data:

Chasen, Craig, 1950–
  Safety manager's guide to office ergonomics / Craig Chasen.
    p. cm.
  Includes index.
  ISBN 978-0-470-25760-9 (cloth)
1. Human engineerign. 2. Work environment. 3. Offices–Health aspects. I. Title.
  T59.77.C485 2009
  658.3'82—dc22

                                                                                2008044562

Printed in the United States of America

10 9 8 7 6 5 4 3 2 1

CONTENTS

PREFACE

The practice of ergonomics is exactly that—a practice—an ongoing activity that engages the principles of human factors, biomechanics, and anthropometry to support and direct the repositioning of a person at work and the components of his or her work environment. The dimension of that repositioning in the real world can be miles away from the examples so regularly shown within common ergonomic guidelines: of a person sitting in perfect, neutral postures. The complexity of office ergonomics does not arise because the optimal end result is new or unknown, but rather by the challenge of what specifically needs to be, or can be, changed to attain that optimal solution. Even if cost restrictions are discounted, converting a workstation that is causing discomfort into a configuration that provides comfort and neutral postures can be very daunting.

Luckily, the days of four-legged chairs and sharp desk corners are for the most part behind us in our current corporate ergonomics enlightenment, and I was hesitant to write this book because I thought that the prevailing trend in ergonomic challenges of the last few decades had been toward solution. But then I was called to conduct an ergonomic evaluation for a longtime employee of an educational institution who had just moved to a brand new, well-funded legal library. I entered the doors confident that this learned environment would be teeming with perceptive ergo-mindedness, reflecting the well-circulated ergonomic knowledge now available.

I arrived early so that I could peek into offices that were probably quite similarly outfitted, in an effort to ascertain what the workstations were like. No expense was spared in the wonderful wood cabinetry and architectural design, and as I observed these lucky employees in their new adjustable chairs with thoughtfully positioned keyboard trays, I saw three or four employees who had moved their mouses from the keyboard tray up onto the desk surface, about 3 in. higher. How could this be? Why was this architectural grace and grandeur tarnished by so many uneven shoulders, silhouetted by the light from their flat-panel monitors?

As I investigated this dilemma, I interrupted a young woman who appeared from behind to be playing an old Hammond organ with two keyboards at different levels. To help relate the essence of my question to this person, I wielded my ever-ready tape measure, and placing the "0" at the left edge of the tray platform and moving to the right, found that the 24-in. mark landed slightly beyond the right edge. A 24-in.-wide keyboard tray! With most keyboards between 19 and 22 in. wide, there was no room left for a standard mouse on the tray—and I returned to writing this book.

That renewed vision of the ever-changing dilemmas that confront the practice of ergonomics is the basis of this work. I take you with me to many of the most representative evaluations that I have conducted so that you can hear what the employees expressed as their particular discomfort and then see the posture that was the cause of or contributor to that complaint. I close each case with what I provided, or recommended, to resolve their discomfort.

The overused phrase "It's not rocket science" is often applied to office ergonomics—and for the most part, rightly. Under that non-Ph.D. umbrella, I use terminology that can be understood by human resources representatives, safety managers, facilities technicians, and especially the employees themselves. There is a glossary at the end of the book that defines briefly a few basic words from the vocabulary of everyday practice because they are effective descriptors and commonly used.

With Dilbert regularly running a cartoon strip involving ergonomics, I had hoped that the dark days of desk posture were behind us, but I have learned that even the most intelligent employees can concoct contortionist poses as they work, and any number of job demands, cost cutbacks, limited space, or just "being too busy" still form fertile ground for pronounced pains in today's workplace. This manual should expedite your working knowledge of office ergonomics and allow you to provide very successful ergonomic interventions.

CRAIG CHASEN

Boulder, Colorado

INTRODUCTION

You have turned to this guide, I trust, because you are already familiar with ergonomics and the impact on employees from ergonomic aspects involving posture, repetitiveness, and/or force. Executing a successful ergonomics program in your workplace has so many potential benefits for the employee that it really rises above the realm of typical safety programs, where it generally resides. What separates ergonomics from other safety programs is the opportunity it presents to stave off an injury by addressing a person's reports of discomfort before that discomfort becomes debilitating or a reportable injury.

The direct benefits of reduced injuries and increased productivity will be quite obvious, but there are also indirect benefits that will surface after the fact, as I learned in my past role as environmental health and safety (EH&S) manager for a midsized corporation in the data storage industry. Those indirect, or secondary, benefits are described in Chapter 1.

The information presented in the book is organized so that when you receive a report of discomfort from an employee or a diagnosis from a doctor, you can turn to a chapter that addresses that body part for a description of the most common causes and solutions. Accordingly, the body parts identified are primarily in the upper extremities. You will also note that every situation depicted, consistent with my experience in today's workplace, involves computer use as a primary cause underlying a report of discomfort.

Safety Manager's Guide to Office Ergonomics, By Craig Chasen
Copyright © 2009 John Wiley & Sons, Inc.

The information provided is culled from over 4000 ergonomic evaluations that I have performed in the 14 years leading to the creation of this book. The recommendations made in the evaluations included here were successful and thus serve as confirmation of the solutions noted. Because I have seen so much duplication in scenarios causing discomfort in office environments, a guide like this can expedite handling of the most pervasive situations. By presenting a wide variety of real-world examples from what I have actually encountered, together with the specific solutions and interventions that were successful for them, users of this guide can replicate proven interventions to fit the situations they have before them.

You will see repeated use of the term *discomfort* to describe any number of more specific complaints or symptoms offered by the people featured in the evaluations. This is intentional, so that the determinations and documentation made by the evaluator, who is often not medically trained, do not misrepresent the actual medical situation at play. Employees will frequently misdiagnose their symptoms, and even using the word *symptoms* can create an escalation that is not appropriate. You should also be aware that it is not uncommon for a medical provider, who is not an occupational specialist, to cite an incorrect diagnosis during the initial development of discomfort, and many will use the vague term *overuse* as their formal diagnosis. It has only been in recent years that many medical professionals have recognized that in additional to treating the symptoms of discomfort, they need to recommend attention to the potential causes of the discomfort, which is where an ergonomic evaluation becomes an invaluable tool.

The premise I work from as an ergonomist is to be a problem solver for the discomfort issues presented by an employee, not to be someone whose goal is to transform that person into an "ergonomic poster child" by imposing one stratified ideal posture template on every person. I see many well-meaning ergonomists leap right into making changes to a workstation setup before they have fully assessed and documented the configuration and postures presented initially. In so doing, they often make modifications that are unnecessary, unwanted, and far from what the focus should be. Such extraneous changes dilute the modifications that would really be pertinent in correcting the discomfort reported, and can actually introduce secondary stressors that cause new problems.

Although this book provides solutions to the most repeated situations that I have observed in 14 years, ergonomics is specific to each person because of his or her size, work activities, and workstation configuration. For that reason, I provide several scenarios for each area of discomfort, so the user of this guide can get a sense of how various conditions and workstation layouts can affect the causes and corrections of the discomfort reported.

I provide details regarding the proper approach to performing an effective ergonomic evaluation in Chapter 3. The evaluation excerpts that comprise the main content of this book are presented consistently in the report format that is recommended.

I have deleted all references to manufacturer-specific products to avoid affiliations with product manufacturers or vendors. As a consultant, I need to be objective in my findings and recommendations and cannot be influenced by anything other than the inherent features of a product to cite it as a potential solution.

By using this book, with its results of more than a decade dedicated to ergonomics, I am confident you will begin to deliver ergonomic interventions that will be exacting and effective and will advance your ergonomics expertise easily and expeditiously.

1

COMPONENTS OF AN EFFECTIVE ERGONOMICS PROGRAM

Working as a consultant for numerous companies and governmental organizations allows me to work within a variety of program structures for their ergonomics program. Those program structures range from companies that bring me in at the request and authorization of one designated employee who manages the program, to companies that allow the employees at large to contact me directly to request an evaluation, with no advance authorization or approval.

Regardless of the program structure, there are four elements in every ergonomics program, whether the ergonomist who performs the evaluation is an employee or a contractor.

1. Employee request process
2. Standardized report form
3. Designation of recipients of evaluation reports
4. Designation of equipment vendors

EMPLOYEE REQUEST PROCESS

One of the foremost decisions that a company must make regarding their ergonomics program is how proactive or reactive they choose to

Safety Manager's Guide to Office Ergonomics, By Craig Chasen
Copyright © 2009 John Wiley & Sons, Inc.

be. While ergonomics-related discomfort and common solutions are becoming more and more visible in today's workplace, many companies do not examine an employee's workstation until their discomfort has progressed to the point that medical attention and/or a Workers' Compensation claim has arisen from it. This is a very inefficient strategy because the results from an ergonomic evaluation of an injured worker's workstation consistently yield recommendations that, in almost every case, will be no different than if the evaluation had been performed when the discomfort was just beginning. Early intervention in ergonomics is the best means to prevent a discomfort report from becoming a Workers' Compensation claim, to avoid Occupational Safety and Health Administration (OSHA) reporting, and to eliminate medical expenses. Accordingly, the sooner that discomfort reports are directed to an ergonomist for resolution, the sooner that recovery can be achieved. When the discomfort is allowed to linger, the measures needed to solve it increase in magnitude.

A reactive approach to discomfort and the injuries that traditionally reside within an employer's safety program should be reexamined to realize the unique potential that exists for injury prevention. Unlike workplace injuries of lacerations, falls, or foreign objects in the eye, ergonomic-related injuries will almost always show early signs of development, thereby allowing intervention before the condition becomes serious. Ergonomics-related discomfort is unique as a workplace injury in that the potential injury that could result will always broadcast warning signs that can be overcome, if those warning signs are acknowledged and addressed early. Additionally, even when an ergonomics-related injury reaches the attention of a medical provider, the provider may not request or suggest an ergonomic evaluation of the employee's work activities, which means that the symptoms will be treated without ever addressing the probable causes.

In contrast to the reactive stance, where the employee is clearly affected, the ways in which companies provide proactive ergonomic interventions vary greatly. Many large companies schedule a very brief evaluation for every new employee, and this is most effective if a company has a very standardized workstation layout that is replicated hundreds of times and the ergonomic "solutions" are also very standardized. Depending on how long a new employee has been at a workstation, he or she may not yet realize how work is affected by the workstation, and new employees may be reluctant to complain.

Many employers require that employees go through an online or computer-based ergonomics training program before an ergonomist is assigned to work with them. The computer training can be a useful tool to help employees learn basic ergonomic principles and optimal workstation configuration, and they may solve their discomfort concerns on their own using that information. If that training does not resolve a discomfort problem, employees are then allowed to request individual ergonomic evaluations. I find that so few employees actually resolve all their issues from computer-based training that I dissuade companies from purchasing elaborate programs, since employees often request personal evaluations anyway. (For more about the benefits of providing personal attention to every employee, see the section "Unforeseen Benefits" later in the chapter.) Accurate ergonomic interventions are very individualistic within their common causes, which is why you will see that the evaluation excerpts presented herein often have similar scenarios but subtle differences.

To request an ergonomic evaluation in most companies, an employee is required to contact the health and safety or human resources department, which then contacts an ergonomist to work with the employee. Sometimes the employee's department is charged for the cost of the evaluation, but that is difficult to structure because evaluation costs will vary depending on the severity of the problem and the extent of the recommendations offered. Many large companies in which the costs for ergonomic evaluations are built into the environmental health and safety (EH&S) program allow employees to request an evaluation from a website. A service request is then transmitted to an in-house ergonomist or an outside consultant without an approval process. Those organizations that contract with an outside professional for ergonomic evaluations provide the consulting ergonomist with independent access to their facilities and often direct access into the company's computer systems. Many will even provide a company computer to the consultant so that the employer can control the hardware being used to access the internal computer networks. This model is a very streamlined process for the department responsible for ergonomic evaluations and often promotes early intervention from a responsive contractor who is motivated to work and produce results.

The common thread within any reporting system is that employees are able to recognize and report their discomfort early in its development, so that the remedies can be sought as quickly as possible. This requires that the employer have a safety program or visible mechanism that educates employees on the symptoms of ergonomics-related disorders and how to seek help in mitigating them. Such requirements are typically the thrust of regulatory agencies and insurers, and employers do their best to comply.

STANDARDIZED REPORT FORM

Standardizing the evaluation report for each company is driven primarily by the company's requirements for organizing and using the data received. The resulting forms cover a wide gamut, from handwritten standardized forms, to a computer template that can pull information from specific fields into a database for tracking. Also, the computerized templates may include dropdown menus that have been populated with the typical posture problems to note and/or the solutions (ergonomic equipment items) that have been preselected for purchase.

Too often, the standardized form is a checklist that includes a finite number of "check boxes" that facilitate quick notation of very general postures or desk configurations and guide the ergonomist through the evaluation process. Such checklists are valuable tools to use during the evaluation so that the evaluee's time is not prolonged by extensive handwriting, but they are seldom satisfactory as the only record of the evaluation. An evaluation checklist should have ample space for notes, and an example of a detailed evaluation worksheet is included here. Examples of evaluation checklists can readily be found on the Internet and from companies that sell "ergonomics packages." Throughout this book, the evaluation report format recommended will include an encompassing narrative that provides a comprehensive description of the setting, postures, findings, and recommendations. We do not delve into the legal ramifications, beyond stating that they are an underlying consideration (the Health Insurance Portability and Accountability Act, for example), just as is any other aspect of an employee safety program. Ergonomics adds a bit of specificity, though.

DESIGNATION OF RECIPIENTS OF EVALUATION REPORTS

In companies where the employees are allowed to request an evaluation without authorization, reports are typically sent directly to them with a copy sent simultaneously to their manager and to the safety department. If workstation modifications are recommended, the evaluator may have a direct link to the facilities department and generate service requests directly. Otherwise, the safety department will direct the modifications, especially since the ultimate decision to make changes or to incur expenditures is a business decision. Some companies have the report (no matter what format) sent only to the person who requested the evaluation, who will review the findings and recommendations before any action is taken and before the report is sent to the employee. Although the practise is rare, in some companies the employee never sees the evaluation report.

DESIGNATION OF EQUIPMENT VENDORS

Small companies and those without an evolved ergonomics program may have no designated vendors for ergonomics-related items other than the office equipment supplier. In recent years office equipment suppliers have been expanding their offerings for ergonomic solutions, but they often do not carry the specialty items that are needed for some ergonomic challenges. If a company does have a relationship with an equipment supplier that carries a more sophisticated line of ergonomic products, the company will have a greater opportunity to acquire solutions that are more closely focused on the specific remedies needed and recommended. Also, ergonomists who are not dedicated to ergonomics often find that suppliers can educate them on new ways to position employees, their tools, and their workstations.

Many companies preselect a finite list of ergonomic solutions, which helps standardize the items for their employees and streamlines the process of making recommendations. However, as an ergonomist becomes more skilled at associating optimal solutions with the variety of discomfort and positioning issues that arise, having access to as many products as possible will increase the evaluator's effectiveness.

UNFORESEEN BENEFITS

Beyond the company impacts and programs outlined above, there are some unforeseen benefits from an ergonomics program that I learned by accident when I implemented my first ergonomics program as an EH&S manager at a large high-tech company in Boulder, Colorado. I found that after I had introduced a proactive ergonomics program that included attending a new employee orientation every week, overall compliance with the company safety program increased dramatically. During the weekly orientation, I invited every new employee to allow me to spend 15 to 20 minutes with him or her at the person's new work area to conduct a cursory examination of how the workstation "fit." Often, I could provide a simple modification that was helpful, or if nothing else, the "safety manager" got to meet independently with the new employee, which would not have happened without the ergonomics component.

The weekly orientation became a standard component of the safety program, with the risk management department noting that employee compliance with safety activities was at an all-time high. Initially, this was attributed to any number of vague notions, such as social awareness, the culture of the community, or the charisma of the current safety manager. However, in the second year of these group presentations and individual visits, an employee survey distributed by the human resources department revealed that those who participated in a brief ergonomic evaluation during their first week of employment saw it as a positive experience within the safety program and became unwitting allies. By spending a short amounts of time with employees learning a bit about what they do (which always arises from examining their work behavior), I became a closer work associate. Therefore, the survey indicated that subsequent requests for safety compliance in other safety-related programs, such as hazcom training or even fire drills, were heeded with much more attention, confidence, and perhaps obligation because of the tangible assistance they received from the safety program.

So not only are ergonomic concerns a hazard you can manage and reverse through early intervention, they can create an amazingly easy way to win sincere support for the safety program as a whole!

2

SAMPLE ERGONOMICS PROGRAM

The sample program we provide in this chapter is a framework model for a midsized company that uses internal resources only. The employees of the internal EH&S department manage the program and conduct the ergonomic evaluations as well as the ergonomics training. The company has employees in office environments and in manufacturing and assembly operations, and is asked to add the pertinent details relating to those environments.

ACME CORPORATION ERGONOMICS PROGRAM

At Acme we strive to maintain a safe, comfortable, and productive work environment for all employees. A key component of achieving that goal is to understand and apply ergonomics to the workstations and activities of our employees. *Ergonomics* is the science of human engineering, which involves designing a workplace that "fits" a person's physical dimensions, with the intent of preventing work-related injuries and illnesses and increasing comfort and productivity. The premise of our program is to provide the resources necessary so that all employees can be made of aware of ergonomic considerations in their work activities and have their workstations configured within established ergonomic guidelines.

Safety Manager's Guide to Office Ergonomics, By Craig Chasen
Copyright © 2009 John Wiley & Sons, Inc.

Responsibility

Responsibility for this program begins with the manager of environmental health and safety (EH&S), who serves as the focal point for information on the causes, concerns, and solutions for ergonomic issues. The EH&S department works with the Workers' Compensation administrator and the facilities department to aid all employees regarding their workstations configurations and the activities involved in performing the tasks required for any job assignment. The EH&S department creates and implements programs to improve awareness of ergonomic issues and provide early intervention for employees who believe that they are developing symptoms related to repetitive motions, awkward postures, or workstation design factors.

Program Elements

1. *Office equipment standards for a new office setup.* The EH&S manager works with facilities personnel to decide which products to purchase for the standard office setup.

2. *Manufacturing and service workstation setup.* Manufacturing employees who believe that their workstations require modifications notify their lead or supervisor, who in turn contacts the process engineer for their area. If the process engineer cannot resolve the issue, an employee contacts the EH&S manager.

3. *Manufacturing tools.* The EH&S department, process engineers, and advanced manufacturing staff establish standards for hand tools that are safe and comfortable while maintaining an acceptable level of productivity.

4. *Training.* The EH&S department conducts ergonomics training programs on an ongoing basis that are offered to all employees. The department updates all training information, including computer-based ergonomic training and computer station setup documentation on a continuing basis. Ergonomics is a topic for every new employee orientation ression, where employees are advised of Acme's ergonomics program and who to contact with any concerns.

5. *Workstation evaluations.* Ergonomic issues are to be reported at the first sign of repetitive strain or continuing upper-extremity aches, pains, or numbness. Every employee has access to the online training programs, which provide a self-assessment of their activities, postures, and workstation configuration. If their concerns are not resolved through the online information, a link is provided to request an ergo-

nomic evaluation. Employees provide their location, phone number, work schedule, and briefly describe their discomfort.

A member of the EH&S department then performs a workstation evaluation at a time that is acceptable to the employee and his or her manager. A written report of the evaluation findings is sent to the manager with recommendations for corrective action as well as notations of any improvement achieved during the evaluation. If the recommendations require expenditures for equipment, management approval is necessary.

6. *Ergonomic equipment recommended.* The evaluator will include in the written report details of all equipment recommendations, and the report will be sent to the employee's manager. Any equipment that requires a purchase request must be approved by the evaluee's manager. If the equipment requires installation or assembly, the evaluator will generate a facilities service request for that work. If equipment items were provided from the EH&S inventory at the time of the evaluation, a charge-back form will be send to the evaluee manager with the report. EH&S department has relationships with prequalified vendors for all recommended ergonomic equipment used at Acme.

7. *Exercise and stretching programs.* Stretching and exercise routines approved by the Acme-designated medical provider are conducted regularly in the manufacturing and service areas. These routines are made available to all employees, and all departments are encouraged to participate in stretching and exercise periods for all employees.

3

THE EVALUATION PROCESS FOR OFFICE ERGONOMICS

Ergonomics is about problem solving. You need to know and understand the impact of all problems identified and the most effective solutions to those problems. This book is a guide to the correlations between the most prevalent discomfort issues and the optimal solutions that are readily available and easily engaged. The ergonomist performing office workstation evaluations is like a detective introduced, perhaps, to an unfamiliar work area, who must quickly recognize and assess all pertinent aspects of an employee's "working world." Like any process, having an organized structure for an ergonomic evaluation will ensure that it is comprehensive as it is conducted, and clear as it is reported. Below, I provide the report structure recommended, which will also serve as a guide to the evaluation process.

THE REPORT STRUCTURE

To keep the evaluation components organized in a logical sequence, I recommend the headings that follow immediately below. The evaluation examples in this book use various combinations of these categories, but they will always fall in the order listed below. I include an example

Safety Manager's Guide to Office Ergonomics, By Craig Chasen
Copyright © 2009 John Wiley & Sons, Inc.

of a complete report at the end of the chapter, which is also included as excerpt 8 in Chapter 5.

Evaluation Objective

Who requested the evaluation? Is this a proactive review, or was it requested by the employee's medical provider or suggested by a physical therapist? Is the evaluation being used to attain a specific goal?

Subjective History

How tall is the person, what is his or her dominant hand, and does the person wear corrective lenses? How long has the employee been at this job and at this desk, and how many hours does he or she work per day? What discomfort does the person report, when did it begin, and which activities aggravate the discomfort?

Observations

This purpose of this section of the report is to clearly document precise measurements of the work surface, chair height, computer components, and so on. Here you record the measurement of angles in pertinent postures and movements, such as wrist angles, head position, or an extended arm reach. Inquire about the length of the person's commute to and from work; and if the person spends a significant amount of time in his or her car, are his or her driving postures awkward or uncomfortable? Does the person have any awkward sleeping postures, such as folded wrists, which could be aggravating the affected body part? The person's activities outside work, including hobbies and recreation, will have an effect, yet those specific activities may have to be excluded from evaluations that are related to Workers' Compensation claims and other employment-specific reviews. Consult your company's risk management or human resources department about restrictions to the information you document and about related programs such as the Health Insurance Portability and Accountability Act (HIPAA).

Direct the employee to engage in germane postures for you to review, such as viewing the very bottom and very top of his or her monitor screen to observe neck postures. Request that the person perform specific actions, such as mouse operation, typing, or reaching to dial a phone. Photos are very effective in making exacting observations, and by using a digital camera the images from an evaluation can be inserted into the report easily and quickly. Having photos of the postures that were the likely contributors to the discomfort are very helpful for medical providers and for anyone not familiar with the employee's workplace. Similarly, photo representations of the modifications made and any postural changes are very helpful to the evaluee. More information on using photos is included in the section "Capturing the Essential Data" later in the chapter.

Corrective Action Objectives

In light of the observations, identify the postures and/or movements that should be the primary and secondary focus for change. An objectives section is not always included in the report, but it can provide a brief outline of the important aspects directly related to resolving the discomfort reported.

Recommendations

The recommendations should be recorded in the form of short statements that briefly describe the modifications made or suggested. This helps the reader scan the recommendations quickly to assess the areas for change and the magnitude of that change. The recommendations should be numbered, listed in the order of importance, and should each identify a specific modification. Accordingly, they should be grouped so that recommendations that influence or affect each other are grouped accordingly. For example, if a special keyboard is being recommended as well a keyboard tray, make sure that they are separate recommendations but listed close to each other. This will help ensure that the keyboard tray configuration can be accommodated by the keyboard you are recommending. If you are recommending products to purchase, be as specific as possible. If you know the vendor that the employer uses for the devices you are recommending (which is always helpful), insert part numbers and/or images that depict the vendor's listing of that item.

Outcome Expected

This section can be included immediately after a specific recommendation so that the recommendation itself can be stated in a short phrase, as in a bullet list. In describing the rationale for the recommendation, you can explain the benefit that will accrue to the posture and/or movements of the evaluee. Using photos in this section is a very effective way to demonstrate the benefits of any modifications made during the evaluation and to help portray the recommendations after the fact.

Summary

A summary section is not common, but can be useful to cite briefly significant aspects of the overall situation and to reinforce the most important recommendations.

CONDUCTING THE EVALUATION

Conducting an ergonomic evaluation in an office environment is a very rewarding process that can be seen as a bit of a game, in which the ergonomist is presented with reports of discomfort and must investigate the situation to uncover the root causes. Here again, the sequence of the process is critical to an effective outcome.

The Initial Encounter

The in-house ergonomist has an advantage over the consultant who is formally escorted to the workstation to be assessed, because an in-house employee can "sneak up" on the person to evaluate and observe without the person knowing that he or she is being observed. This "catching them in the act" will often reveal the actual working postures more accurately, since employees tend to "pose" for the ergonomist. Getting a person to relax and assume "real" positions takes a bit of psychology and a graceful "deskside manner."

You begin by gathering information that "paints a picture" of the employee and the setting using a worksheet for your notes. (An example of an evaluation worksheet is included at the end of the chapter.) Height, dominant hand, job title, and how long he or she has worked at the current desk are the foremost questions to ask. Then, as you inquire about the discomfort, when it started, and how it affects him or her, you will find the person to be very forthright in explaining where it resides and why the employee or medical provider or therapist requested the evaluation.

This is also the evaluator's opportunity to gauge the motivations of the evaluee. It is not uncommon for an employee with very little discomfort to request an evaluation if a co-worker recently received a new chair or other slick new device for his or her workstation. For example, wrist discomfort can appear to be a contagious malady when you find that a neighbor of an employee who received a specialized mouse suddenly develops discomfort that is identical to that reported by the co-worker. This is one reason that the subjective history is an important area to discuss and document. Also, in keeping with the approach that I recommend, we are performing the evaluation to resolve discomfort issues, not just to reposition the person to resemble the "perfect posture" model that so many simplified ergonomic posters portray.

Being clear in understanding and documenting employee reports of discomfort is also very helpful when you work with an employee who really has no awkward postures and whose workstation almost typifies the ergonomic ideal. When that occurs, the ergonomist needs to recognize that something else is at play, and I recommend reviewing Chapter 7 to gain a sense of how to work through such unavoidable encounters.

Also, many companies request an ergonomic evaluation for every new employee: with or without reports of discomfort. In those situations, you have to be careful that (a) the employee does not discount actual discomfort because he or she does not want to appear to be demanding, or (b) the employee thinks that this is an invitation to get new "toys" or "creature comfort" items that are not really needed.

After you have logged the reasons that the evaluation has been scheduled and noted the employee's own assessment of the current discomfort, you are ready to enter the active stages of an office evaluation. Your work at this point requires nothing more than your astute observational skills and some very simple tools: a small tape measure, a digital camera, and the worksheet used for note-taking.

Capturing the Essential Data

You now ask the employee being evaluated to start working in his or her typical manner and to try to disregard the ergonomist, who is taking measurements with a small tape measure and photographs as the employee works. As you proceed, it is often helpful to ask some non-intrusive questions to distract the person from feeling your visual scrutiny and your proximity as you measure chair height, desk height, and so on. I often ask such things as if he or she likes the particular keyboard or mouse being used or what type of workstation he or she had at a previous job: mostly small talk, but you will often learn that the positions of the chair, keyboard, or monitor are merely "the way they were" when the employee moved into that workspace. It is surprising how often employees will not modify a newly acquired workstation or take the time to adjust the chair. In this phase you are looking to identify the postures and motions that are related to the discomfort while you document the initial configuration of the workspace, without making *any* changes.

When you take your initial configuration photos, try to remember the position you were in and how the image was framed, so that you can use the same framing after modifications have been made to depict the changes clearly in a "before and after" comparison. The photos will also help you remember what you saw. Sometimes my notes are not as inclusive as I would like, because I'm talking with the client while making the notes and I don't want to take the client's time to make detailed notations. Photos often remind me of things such as what kind of keyboard was used, and I have even used photos after the fact to help me "measure" the depth of a work surface. As I review the photos and see few 8.5- by 11-in. paper lying on the desktop, I can make a close approximation of the desk size by comparison. There are a number of similar assessments that can be made from the information revealed in the photos. This is especially true for long-distance evaluations, as noted in Chapter 6. The photos in the report also help the employees see their own posture. Employees very often do not recognize the irregular postures they are in as they do their work. Neck postures are often the most revealing, because people cannot see their own neck positions.

It is helpful to be knowledgeable in working with digital photos so you can ensure that they display well in a printed version as well as on a computer screen. You can enhance the brightness of an image or crop an image to display exactly what you want to portray in your report. In the last four years, every report that I submitted was delivered electronically as an e-mail attachment. The reports are created in a typical word-processing program and then distributed as a PDF (portable document format) file. As digital cameras evolve, be careful to monitor the file size of photos you use. When delivering a report by e-mail, it is easy to have photos embedded in the report that make the file size so large that it will be "blocked" by the e-mail servers of your customer.

Beyond a camera, there are many secondary tools that can be useful, such as a simple goniometer, a light meter, a portable scale, or even force sensors. However, for basic office evaluations, where ergonomics is just one of the duties performed by the ergonomist, a tape measure and camera are the only tools required. Once again, as you are "capturing the essential data," whether as a co-worker or as a consultant, it is very important to make all the "initial observations and measurements" before making any changes. It is very easy for enthusiastic and well-meaning ergonomists to make modifications as soon as they observe them. They may observe a clearly egregious posture that calls for an adjustment to "make it right," but this is not the time to modify anything.

For example, the subject's chair might be much too low, and the evaluator will merely raise it to a "better" level. But this change was better for what aspect? It is likely that the chair adjustment was done to place the lower extremities in a neutral posture; however, the chair-height modification will affect the wrist position relative to the desk, and the head position relative to the monitor.

No matter what the reference point was for raising the chair, by changing it you have modified all related postural aspects. If you did not make prior notes about the measurements or positions for related aspects, you have lost your baseline posture, and subsequent modifications may only be necessary now because of that chair modification. The other aspects cannot be measured or assessed properly without returning the chair to its initial position.

On the whole (literally), you must acknowledge that each adjustment you make for one body part affects other body parts relative to their postural relationship to some aspect of the workstation. Until you have a lot of experience, you may not realize the subsequent impact of premature adjustments, and employees themselves will often make "on the fly" modifications, which you should discourage until the initial configuration is captured and documented.

A common example of cause and effect is when a work surface is raised or lowered to attain a better keyboard height, and it subsequently alters the position of the monitor height relative to the user's eye level. For example, people wearing bifocals may have their keyboards very low, with their wrists bent back because their wrists are below their seated elbow height. If a well-meaning ergonomist raises the work surface (or lowers the chair) to place a person's wrists in a neutral posture, a detrimental neck posture may be caused by making the evaluee tilt his or her head back to view the monitor, since it will be higher. If the monitor, in this scenario, is placed directly on the work surface, it cannot be lowered. Therefore, solving the problem of wrist extension from a low keyboard will have incurred neck flexion because now the monitor is too high for viewing through the lower lens of bifocals while maintaining a neutral head and neck posture.

INTERPRETING THE DATA AND IDENTIFYING THE SOLUTIONS

With the pertinent data now collected in notes and in camera, you are poised to put forth modifications as you work with the employee and/or in recommendations in your report. Accordingly, it is now time to retrieve the knowledge gained from a catalog of successful interventions that comprise the core of this volume. The examples in the book demonstrate the process outlined above, and the "problems" noted will be coupled directly to the optimal solutions that were proven to resolve them. Beyond the guidelines I have noted here, gaining an education and training in body mechanics and anatomy will enhance your skills in analyzing physical stressors. Still, for the EH&S employee who wants to perform in-house evaluations, these strategies will allow you to progress rapidly with constructive, beneficial results.

SAMPLE EVALUATION REPORT

The Chasen Group

ERGONOMIC EVALUATION REPORT
09-15-08

CLIENT:	**Wanda Dunn**
EMPLOYER:	**Acme**
LOCATION:	**1234 Coyote Drive**
PHONE:	**(303) 555-1234**
DATE OF EVALUATION:	**9-17-08**
JOB-SITE EXAMINER:	**Craig Chasen, CEES**

History

Wanda Dunn is right-handed, 5 ft 4 in. tall and has worked as a paralegal at her current workstation for one year. For the last two months she has been experiencing tingling in her hands, which she reports is especially pronounced when she is editing hardcopy documents. Additionally, she has been experiencing discomfort along the left side of her neck.

Observations

She works at a rectangular desk that is 28 in. tall, with her keyboard and mouse placed on the desk surface and no wrist rest for her keyboard. She has a large "pillow-style" wrist rest for her mouse that she says is very comfortable.

She sits in a small semiadjustable typing chair that places her hips just above her knees, which is optimal, and it has a small backrest at its lowest position. This backrest position does not provide upper back support, and the seat height places her elbows significantly lower than her wrists as she operates her keyboard and mouse. Since her chair does not have armrests, she places some upper extremity weight on her wrists as she rests them against the sharp corner of the desk surface.

Wrists are much higher than her seated elbow height, which is not advised.

Wrist height and no wrist rests create significant contact pressure against corner edge of desk.

Recommendation 1

Install an articulating keyboard tray and move the desk forward, closer to the door.

Outcome Expected   To allow her to position her wrists at or slightly below her seated elbow height, she would benefit from placing her keyboard and mouse on an articulating keyboard tray. The width between the side cabinets of her desk is 27.5 in. and there is room to install a full platform (11 in. × 27 in.) keyboard tray. When using the tray, her chair will be moved about 12 in. back from the desk, so she should move her desk forward. There is sufficient space in her office for the desk to be moved forward 1 ft.

Recommendation 2

Replace the small chair with a fully adjustable task chair.

Outcome Expected   During the evaluation, her backrest was raised from its lowest position to its highest position to allow her to receive support to her upper back and allow her muscles to relax as she sits upright. The adjustment that holds the backrest in place will often lose its grip, and she is advised to check it from time to time because the backrest will slowly move down. Because the backrest will not maintain its height and is very small, she is a candidate for a chair with a taller, more adjustable backrest.

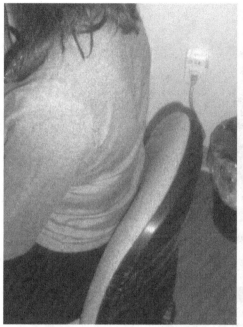 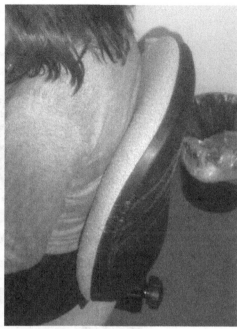

| No middle or upper back support from the small backrest. | Even with backrest raised fully there is minimal support. |

Recommendation 3

Use a document holder between the keyboard and the monitor.

Outcome Expected With the keyboard tray in place, there will be space on the desk for documents directly in front of her. Rather than placing documents on the flat desk surface and to the side, she should use a wide document holder between the keyboard and the monitor. Most document holders designed for this application will drop the lower edge below desk height so that the documents will not block any part of the screen data.

Looking left and down for hardcopy documents.

Neck remains neutral with documents raised and centered.

If you have any questions about this report, please do not hesitate to contact me.

Craig Chasen, CEES, job-site evaluator.

The Chasen Group 2008

EVALUATION WORKSHEET

Name: _____   Date: _____   Phone: _____

Company name: _____   Location: _____

                Bldg.: _____   Room No.: _____   Mail stop:_____

Supervisor_____   Phone: _____

Job title: _____ Time w/company: _____ At current workstation: _____

Dominant hand: R   L   Time at computer: _____ % of day   Mouse-intensive: _____

Work schedule: _____ hours/day _____ days/week _____ hours/week Time off: _____

Age: _____ Height: _____ Corrective lens: Y N   Bifocals   Contacts   Progressive lens

Symptoms/Injury: Diagnosis: _____   Doctor: _____

_____

_____

_____

_____

Date discomfort began: _____   ER: Y N   Time off: _____

Physical therapy: _____   Medical treatment: _____

Exercises: Y   N   _____ x/day _____ x/week   Better   Worse   Unchanged

Job Responsibilities:

_____

_____

_____

_____

_____

_____

Lifting: _____ lb _____ x/day   Other: _____

EVALUATION WORKSHEET	page 2

Workspace dimensions: _____ Main work surface dimensions: _____

Work surface height: _____ Keyboard tray: Y N   Tray positioning: _____

Keyboard type: _____ Wrist rest: Y N   Mouse type: _____ Wrist rest: Y   N

Monitor position: _____ Monitor height: _____ Monitor distance > eyes: _____

Wrist extension L: _____ Wrist extension R: _____   Neck: Neutral   Flexed   Extended

Ulnar deviation L: _____ Ulnar deviation R: _____   R shoulder: Neutral   Flexed   Abducted

Elbow extension L: _____ Elbow extension R: _____   L shoulder: Neutral   Flexed   Abducted

Chair type: _____ Seated elbow height: _____   Hips are: above/below knees

Seat pan: height _____ depth _____   Armrests: _____ adjustable _____ width _____

Back support: _____ Footrest: Y   N

Computers being used   1st: _____ 2nd: _____   KVM: Y   N

Corrective Actions:

1 _____

_____

2 _____

_____

3 _____

_____

4 _____

_____

5 _____

_____

6 _____

4

AREAS OF DISCOMFORT

Discomfort resulting from the ergonomics-related activities of workers in an office environment will almost always present in their upper extremities, with occasional hip and leg complaints. In this chapter we outline how employees typically relate their discomfort to the ergonomist, how that body part is typically affected, and the most common causes. Since the audience for this book is not expected to have extensive knowledge of medical or anthropometric disciplines, the discomfort descriptions are related in the same terminology that will typically be expressed by the employees who are being evaluated.

- Shoulders
- Elbows
- Wrists
- Forearms
- Lower back
- Upper back
- Neck
- Eyes
- Hips

SHOULDER DISCOMFORT

Shoulder discomfort is the most frequent area of discomfort reported in my evaluations of employees who work in any computer-based activity. Typically, it presents on the side of their dominate hand; however, it may manifest on their nondominant side. Even when the causal activity or posture exists on the dominant side, the resulting discomfort may present on the opposite side. The reverse is also true.

The most common causes are:

1. *Computer work with wrists higher than seated elbow height.* This posture is, by far, the most common cause of any computer-related upper extremity discomfort. Interestingly, people seem to sense that their elbows should be at wrist height as they work, and they will unknowingly engage two primary postures to raise their elbows to wrist height when their seated elbow height is lower than the height of their keyboard and mouse. Those two postures are:

A. The user will move the keyboard back from the desk edge so that his or her arms must be extended forward to type, which moves the elbows forward of the torso. This raises the elbow, which can attain wrist height, especially when the user is resting his or her forearms on the desk surface.

B. The user will move his or her elbows outward, away from the torso, and place them on the desk surface to either side of the keyboard and mouse. This is done most frequently when the user is seated at a corner-oriented workstation which has work surfaces that angle toward the user. A typical version of this is where the person's nonmousing arm is placed on the desk to the side or in front of the keyboard, with the mousing arm fully extended for mouse movement.

In these two postures, the user's shoulders are rounded or abducted, respectively, and he or she will not be able to position the upper back against the chair's backrest. Again, this is the most common single cause of upper extremity discomfort and can be found in almost any work environment.

2. *User keyboard on a keyboard tray below the desk surface, with the mouse placed and used on the desk surface.* This situation is found most frequently in older furniture, designed (before Windows) with an adjustable section of the desk only large enough for the keyboard. Typically,

these keyboard platforms are only 19 to 23 in. wide, which does not leave room for the mouse. The user will then place the mouse on the desk (left or right), which is often 2 to 4 in. higher than the keyboard surface.

3. *Repeated movement of the hand between their mouse device and the home keys of the keyboard.* This is more prevalent on a person's right side, since most keyboards have the numerical keypad and arrow keys to the right of the primary key set, which increases the distance from the mouse device. Occasionally, a bridge mouse platform is placed above the numerical keys to bring the mouse closer, but such bridges elevate the mouse an inch or more above the keyboard and create an awkward posture. Accordingly, the "mouse bridge" introduces the problematic postures mentioned in example 2.

4. *Working in a chair that does not provide support to the middle or upper back while in a working posture.* The absence of contact to the middle and upper back while performing computer work will frequently cause a person to slouch forward and direct upper torso support to the elbows, arms, or wrists. Introducing upper back support from the chair backrest while the person is in an upright working posture usually has a pronounced benefit in improved posture, allows the user's muscles to relax, and reduces fatigue.

5. *Cradling a phone receiver in the neck and shoulder, especially as other activities are performed with that arm.* Computer use in concert with frequent or extended phone conversations while the phone is cradled is a ubiquitous posture that places a significant physical demand on the shoulder and neck. This posture, coupled with stress or even cold air from an air-conditioning vent, can be damaging to the shoulder on either side of the cradling. The common stresses that affect employees will often reside in the shoulder area, and the discomfort will be compounded when the shoulder and/or upper back are in an awkward or unsupported posture.

ELBOW DISCOMFORT

Elbow discomfort is most frequently diagnosed as "tennis elbow," epicondylitis, or tendonitis. It is typically recurring pain on the outside of the upper forearm just below the bend of the elbow, and occasionally radiates down the arm toward the wrist. When the discomfort is felt on the inside of the elbow, on or around the joint's bony prominence, it is often called "golfer's elbow." Elbow discomfort can also be bursitis, arthritis, or a more serious condition.

In the most common, tennis elbow, condition, movements such as gripping, lifting, and carrying aggravate the discomfort, even for light-weight objects. The discomfort can continue for as little as three weeks or as long as several months, and it can take a long time to resolve if the computer activities and postures that contribute to it are not resolved.

A common situation for elbow discomfort is caused by the compression or "entrapment" of the ulnar nerve. The ulnar nerve travels around the elbow and functions to give sensation to the little finger and the half of the ring finger, next to the little finger. It also controls most of the little muscles in the hand that help with fine movements and some of the bigger muscles in the forearm that help to make a strong grip. When this ulnar nerve entrapment arises, the person will typically feel numbness and/or tingling in the little and ring fingers of the affected hand.

Striking the ulnar nerve creates a sensation in the arm known as "hitting your funny bone." This nerve may travel around the elbow in different positions for different people and is often surgically relocated toward the front of the elbow (ulnar nerve transposition) to reduce the potential of compression or the degree that it is pulled as the elbow is flexed.

The most common causes are:

1. *Contact pressure from resting the elbow on a hard surface for prolonged periods.* The most common cause of elbow discomfort in computer users is resting their forearms on a hard desk or bending an arm so that the tip of the elbow is placed on a hard surface as it supports the arm and shoulder. It is a frequent posture in call centers where employees have limited activities during long phone calls, and they tend to lean on their elbows if their hands are not occupied.

2. *Repeated movement of the hand between the mousing device and the home keys of the keyboard.* This motion can be damaging to the elbow or shoulder and will affect the elbow most often if the elbow is "anchored" on an armrest, or positioned as a pivot point for that movement.

3. *Working or sleeping with the arm in a tight bend where the wrist is close to the shoulder.* A tight bend at the elbow in a static or active posture can affect the elbow as well as the ulnar nerve that travels around the elbow.

WRIST DISCOMFORT

Wrist discomfort is the most commonly recognized area of discomfort in computer users, largely because of the media exposure surrounding carpal tunnel syndrome (CTS). However, CTS is frequently attributed to wrist injuries that are not actually CTS, because of the complexity of the composition of the wrist. Computer use as a singular direct cause of CTS has not been proven, but its prevalence in computer users and the symptom relief from ergonomic interventions underlie the findings and solutions contained in this material.

Wrist discomfort can manifest in a number of forms that arise during the course of a person's work or when the person's wrists are in a mode of complete nonactivity. One of the most common indicators of wrist problems is when a persons's hands go numb in the middle of the night. I work frequently with computer users who report that they are awoken from sleep because their hands go numb.

The following are the most common causes of wrist discomfort:

1. *Working with the wrists in awkward positions.* Wrist movements are necessary and unavoidable in computer work, but the degree of the wrist angles should be limited. Excessive wrist angles of extension, flexion, and ulnar deviation increase wrist strain, especially when force and/or repetition are added. The most neutral posture for the wrist is a "handshake" posture. Troublesome angles can be exaggerated in people who have hypermobility (often called double-jointedness) in their wrists. Many computer users have varying degrees of hypermobility.

2. *Resting the underside of the wrist on a hard surface during computer work, creating contact pressure (tissue compression).* It is very common for computer users to work with their wrists resting on a desk surface in front of a keyboard, mouse, or calculator. This issue is increasing for laptop users, who often rest their wrists on a hard chassis surface in front of the keyboard.

3. *Repetitive lateral wrist motions at the wrist joint with the forearm anchored.* People often rest a section of forearm on the desk or on a wrist rest in front of a mouse and keep the forearm stationary (anchored) as they move the mouse. This anchored posture of the forearm directs all mousing movement to the wrist joint. This adds a greater physical demand on the wrist than does allowing the forearm to move and thereby reduce some wrist deviations.

4. *Shoulder posture that impinges the median nerve as it travels through the shoulder from the spine.* It is not uncommon for an awkward

shoulder posture to impinge the median nerve, causing discomfort in the hand and inflammation in the wrist.

5. *Sleeping with wrists folded.* It is very common for people to sleep with their wrists in a folded (flexed) posture. This posture, especially when prolonged in duration, puts damaging pressure on the nerves. Physicians often prescribe that wrist splints be worn during sleep.

FOREARM DISCOMFORT

Forearm discomfort is often a secondary symptom of wrist or elbow discomfort, and information provided regarding wrists and elbows will often overlap into forearm discomfort issues. The postures that are specific to forearm discomfort are either contact pressure from resting the forearms on hard surfaces or from extended reaching. These postures can be combined for support, and when they are, it is likely that the user is placing some weight on the forearms.

The most common causes are:

1. *Resting or leaning on forearms against a hard surface, creating contact pressure (tissue compression).* Any number of workstation configurations can create a posture in which the user's forearms are resting on a hard, noncushioned surface. In situations where the wrist is higher than the elbow, the shoulders may be forward, with torso weight delivered to the area of contact. Adding weight to the contact pressure, against a hard surface, increases tissue compression at the area of contact and increases the potential for tissue damage.

2. *Computer work with the wrists higher than seated elbow height.* Even if the forearm is not resting against a hard surface, working on a computer with wrists higher than seated elbow height can add a physical stressor to the forearms.

3. *Moving the mouse hand (primarily right) between the mouse and the home keys.* This is more prevalent on a person's right side because most keyboards have the numerical keypad and arrow keys to the right of the primary key set, which increases the distance from a right-side mouse device. Occasionally, a "mouse bridge" platform is placed over the numerical keys to bring the mouse closer, but such bridges create a mousing surface that is more than an inch higher than the keyboard surface. Therefore, operating the mouse on the bridge surface elevates the wrist from keyboard height and probably above seated elbow height, which is not recommended.

LOWER BACK DISCOMFORT

Lower back discomfort is the most ubiquitous complaint within ergonomics, much as it is in the U.S. population at large. It is the consequence of many diverse causations beyond ergonomics, but sitting is
often the posture that is most problematic for persons suffering from
lower back discomfort. Sitting moves the spine away from its normal
center of gravity and increases the physical burden on the discs compared to a standing posture.

The safety manager who has limited familiarity with anatomy should
know that the lumbar spine is composed of five vertebrae labeled L1
through L5. The locations affected by lower back discomfort are usually
the intervertebral discs between the vertebrae. The disc between L4
and L5 is the primary disc affected. The next most common area of
lower back discomfort is the joint of the lumbar spine and the sacrum
(below the lumbar spine), typically referred to as S1. Accordingly,
employees who have received medical attention for lower back discomfort will frequently report that their discomfort source is at L4/L5, or
L5/S1, or both. L3 is the next most common area to be affected, and
some people have an additional lumbar vertebra, L6. The issue at L4/L5
is that the discs between and adjacent to them are easily made to bulge
at the back (posterior) and can press on the nerves that emanate from
that section of the spine. Those nerves radiate into the legs and will
often cause numbness in the legs, known as radiculopathy or sciatica.
The causes of lower back pain are numerous, and the ergonomic issues
are typically the consequences more than the cause, but poor posture
when sitting at a desk or computer can exacerbate the condition
significantly.

The most common causes are:

1. *Having the chair backrest unlocked so that it will recline easily.* When
a person performs computer work while seated in a chair that does not
have firm back support from the backrest, they must engage their own
muscles to sit in an upright posture. That muscle engagement will introduce muscle fatigue sooner than if the backrest were locked, allowing
their muscles to relax since the chair's backrest is keeping their back
supported in an upright working posture.

2. *Sitting in a chair that allows the lumbar spine to "flatten" where the
natural curve is not supported.* This is typically caused by chairs that
have a very flat backrest without a protrusion that is aligned with the
inward curve of the user's lumbar spine. That curve is called a lordosis
and is referred to as a lumbar lordosis. This can also arise when a chair

has a pronounced lumbar support but the protrusion is adjusted too low, so that it is aligned with the user's hips, below their lumbar lordosis. This causes the hips to be moved forward by the protrusion so that the inward curve of the low back becomes aligned with the backrest section that is recessed above the protrusion, and a significant gap is created.

3. *Using a chair with a backrest that does not reach up to the shoulder blades.* If the backrest only provides support to the lower areas of a user's back, the user will often hunch forward with the upper back rounded and place weight on the forearms. Alternatively, the person will attain a posture where he or she moves the hips forward on the seat pan in order to drop the shoulders so they are in contact with the backrest, thereby receiving shoulder support from the backrest. This posture leaves the lumbar spine unsupported in a position that invites flattening of the inward curve.

4. *Twisting the torso to move the shoulders away from the direction in which the legs are pointed.* This often occurs when the entry to a person's work area is more than 90° away from the forward view of the monitor. If the employee turns frequently to converse with a co-worker at the entry to the cube or office while having the hands on the keyboard, the lower back is subject to an awkward motion that can be very detrimental.

5. *Sitting with the hips lower than knee height.* This will stretch the hamstrings and can flatten the lumbar spine, putting pressure on the front of the intervertebral discs.

6. *Having a seat pan that is too shallow.* When a limited area of the upper legs is being supported by the chair seat pan, the user's weight is distributed to a smaller area, increasing the load on the small area being supported. Although the immediate impact is to the user's legs, the back is also affected.

7. *Bending down for files or items positioned close to the floor using poor lifting mechanics.* Although the safety generalist or entry-level ergonomist may not be entirely conversant with some of the principles herein, they are probably familiar with basic lifting mechanics and how poor lifting motions can cause significant back discomfort.

8. *Moving a task chair across a floor surface (typically, carpet) on which it does not move easily.* When a chair does not roll across a surface easily, the user will push and pull, in the chair, using his or her legs and/or arms by grasping the work surfaces. This becomes exaggerated, of course, if the chair does not have wheels, which all task chairs should have.

9. *Lifting or moving heavy office items such as CRT monitors or chairs.* Again, proper lifting mechanics need to be observed.

10. *Working with the wrists higher than the seated elbow height.* This was described earlier as a common cause of discomfort in the arms and shoulders, but it also causes lower back discomfort. That is because a person working with the wrists above their seated elbow height will typically react by moving the shoulders forward. When the person has the shoulders forward of the hips and move away from the chair back-rest, he or she is more easily subject to flattening the curve of the lumbar spine, putting pressure on the intervertebral discs.

11. *Sitting with a wallet in a back pocket.* This is most common for men, and understandably, misaligns the hips by putting uneven pressure on the pelvis and spine.

UPPER BACK DISCOMFORT

Upper back discomfort is less frequent than lower back discomfort and is often connected with shoulder discomfort.

The most common causes are:

1. *No upper back support from the chair.* Like shoulder discomfort, one of the primary contributors to upper back discomfort is having no support from the chair, so that the user arches forward in a hunched posture. Receiving upper back support from a chair does not merely mean a chair with a tall backrest, however. The vast majority of tall-back task chairs available today have the uppermost section of the backrest angle rearward, away from the user's torso. That backrest configuration defeats the purpose of having an upper backrest section unless the user leans back in some type of "executive cogitating" posture. Computer users need the backrest to help them remain very vertical, providing upper back support as they sit in an upright position for computer work.

2. *Having the monitor positioned too high or too low.* Having the monitor raised so that a computer user's head is in a neutral posture as he or she views it is very important. Typically, the user's horizontal line of sight should meet the screen one-third of the way down from the top of the screen. However; in ergonomics, everyone is different, so make sure to test neck posture when positioning a monitor and don't use a common standard for everyone.

3. *Having the chair backrest unlocked so that it will recline easily.* When people perform computer work while seated in a chair that does not

have firm back support from the backrest, they must engage their own muscles to sit in an upright posture. That muscle engagement will introduce muscle fatigue sooner than if the backrest were locked, allowing their muscles to relax since the chair's backrest is keeping the back supported in an upright working posture.

4. *Hunching forward while working.* Working with the forearms on the desk with the head forward of the lower back, and possibly lowered, is an all-too-common computer posture. Although this can be caused by the monitor being low and too far back, it is caused more frequently by the user's wrists being higher than the seated elbow height as he or she types and uses the mouse. This positioning typically has the shoulder flexed where the shoulders are higher than the they would be in a relaxed, neutral posture. Having the wrists higher than the seated elbow height also tends to draw the shoulders forward, which does not conform to conventional wisdom, but it occurs very often. The common stresses that affect employees often reside in the shoulder area and compound the discomfort when the shoulder and/or upper back is in an awkward posture.

NECK DISCOMFORT

Neck discomfort is another common discomfort issue that is usually caused by the position and/or movement of the user's head and neck in order to view the monitor, especially if the person wears corrective lenses. Other postural causes of neck discomfort in a typical office environment include viewing reference documents on the desk surface, turning the neck to view visitors at the doorway, and cradling the phone receiver in the neck and shoulder.

The most common causes are:

1. *The monitor being too high or low.* Although my experience and valued information from Professor Alan Hedge at Cornell University find that the user's horizontal line of sight should meet the screen one-third of the way down from the top of the screen, everyone is different. To determine the optimal height of a monitor screen for a specific user, I take a pen or pencil and point it at sections of the screen while observing the user's neck posture and eye movements. First, I point to an icon or small image at the very top of the screen and ask the person to identify it. Then I do the same thing for an image at the bottom of the screen. (Asking a person to follow the mouse cursor as it is moved to the top and bottom of the screen also works to demonstrate the user's

neck postures at the extremes.) Seeing the neck posture at the upper and lower extremes provides a basis for whether to raise or lower the screen so the user's neck is generally neutral. For most office applications, the "active area" of the screen and the menu items used most frequently are in the upper section of the screen.

2. *Viewing reference documents on the desk surface.* The most accessible space for reference documents that need to be compared to the screen is on the desk to the left or right of the keyboard. Moving the head and the field of vision between the documents and the monitor screen affects the neck because of the motions involved and by vision and light quality. Employees will also place documents in front of the keyboard and view them with the head bend down significantly while extending the arms forward to reach the keyboard, behind the documents.

3. *Wearing glasses that cause the wearer to position the head.* It is not uncommon for a new wearer of bifocals or progressive lenses to tilt the head back unknowingly as they view the screen through the lower lens. This problem can be difficult to solve because the monitor base may prevent the screen from going low enough. However, on CRT monitors it is typically quite easy to remove the swivel base, and since most flat-panel monitors have VESA standard mounting holes on the back, the standing base can be replaced with an articulating arm that will allow the screen frame to go down to the desktop.

Also, employees who wear bifocals and progressive lens glasses, especially those with small lenses, will move their head in various positions so that the proper lens area is positioned for the section of the screen they are viewing. This precise positioning requires them to tighten their neck muscles to maintain and to move their head for the visibility they need.

In contrast, you will find that employees who have played a brass or reed instrument in an musical ensemble using sheet music and a conductor will move their eyes while keeping their head in a fixed position. That is a result of having to maintain a stable embouchure for the instrument while reading the music on a music stand and looking up at the conductor without compromising that embouchure.

4. *Looking down to the keyboard to see the keys, with the keyboard low.* Many computer users have very limited typing skills, and as a result, they tilt their heads downward to identify the keys they wish to use. This posture can be extreme, and positioning the keyboard at seated elbow height, as recommended, may increase the head tilt compared to using the keyboard in a slightly higher position. This challenge

is not easy to solve if the person is unable to "type by touch." A compromise is typically best, where the keyboard is raised above elbow height and is angled down at the front so that the user's wrists are kept flat as he or she types on the elevated keys.

5. *Cradling a phone receiver between the shoulder and the neck.* This is extremely common, and cell phones make it particularly awkward. Even the cushions that can be attached to the receiver should be avoided, since they reduce the physical strain only slightly.

6. *Turning more than 90° from the computer screen to view the entry to the work area.* Where a computer user's doorway is more than 90° to the side of the computer monitor, and he or she frequently has co-workers show up in the doorway to pose a quick question, the computer operator will often turn the neck and upper torso while keeping the hands at the keyboard or mouse position, creating a strained posture in the neck. Other circumstances can also create such a neck turn, and the greater the turn is, the more likely it is that neck discomfort will result.

EYE FATIGUE

The office worker in almost any capacity spends significant amounts of time viewing a computer monitor of some sort. Eye strain is often caused by prolonged periods of time viewing the screen, excessively bright light from outdoor sunlight coming in through a window, or from harsh interior lighting. When viewing a computer monitor, people blink less than half as often as in normal activities, which reduces the lubrication that the eye receives. It is also common for headaches to arise from the situations that cause eye fatigue.

The most common causes are:

1. *Having light sources in close proximity to the monitor screen.* The positioning of a computer station often disregards the location and intensity of the lighting sources for ambient room lighting and overhead workspace lighting. In simplistic terms, a light source that is directly above a monitor or within the field of vision while viewing a monitor, will generally "compete" with the information sent to the eye and cause strain. Having a bare light source that is not diffused, recessed, or lighting the area in the field of vision indirectly creates a distraction that is difficult for the eye to shield while still trying to view the monitor screen. Generally, the level of ambient lighting in the area should be lower than that typically found in office environments.

2. *Seeing the reflection of another light source on the surface of the monitor screen.* This is a form of glare, where the screen surface reflects light that is behind the person viewing the monitor such that it is bright enough and positioned to reveal an image of the light to that person. Typically, just repositioning the screen so that the image of the extraneous light source is not seen is all that is required. However, this is not always easy to achieve. Although there are many varying recommendations on monitor positions that place the monitor well below the user's horizontal line of sight, I have found that such low positioning is more likely to reflect overhead lights from behind than if the monitor is elevated.

3. *Having the monitor's refresh rate set at 60 Hz, especially for CRT monitors.* Because fluorescent lights flash on and off 60 times a second, and most monitors refresh the screen image 60 times a second, there is a visual phenomenon that is not easily perceptible but can be disturbing. The screen image on a monitor (especially a CRT monitor) will often demonstrate a pulsing screen image that is most detectable on wide areas of a white or light-colored screen surface. The refresh rate should be modified to an odd number that is far from 60. Often 75 or 85 Hz is available, and 85 Hz is preferred because it is furthest from 60 Hz. This does not completely remove the flickering phenomenon but reduces its perceptibility.

4. *Prolonged viewing of the monitor screen without focusing the eye on distant objects.* When the eyes do not have an opportunity to focus on objects farther than the immediate walls of a small work area, they do not get exercised. Therefore, when the eyes try to focus on a distant object for the first time after a long computer session, there is additional strain from the lack of exercise.

HIP DISCOMFORT

Discomfort at the hips is an infrequent complaint related to office ergonomics, but like all such discomfort issues, it can be a component of cumulative trauma involving a combination of causes. Typically, it presents along the sides of the hips and is most pronounced when a person is getting into, or up from, a chair.

The most common causes are:

1. *Sitting with the hips significantly lower than knee height.*

2. *Sitting in a chair that has armrests so inwardly close that they press against the sides of the hips.*

3. *Sitting in a chair with the seat pan angled down in the front so that the user must place some of his or her weight on the legs to stay positioned in the chair.* This seat pan position is often accidental, where the user does not realize that the seat pan can be angled back, into a horizontal orientation. Many fully adjustable task chairs have an adjustment lever that prevents the seat pan from being angled forward.

4. *Sitting in a chair that has a seat pan which is very "cupped" from side to side and/or the side edges of the seat pan having a hard, raised edge.*

5. *Sitting with a wallet in a back pocket.* Most common for men, this misaligns the hips by putting uneven pressure on the pelvis and spine.

5

EVALUATION EXCERPTS

This chapter comprises 38 actual ergonomic evaluations that were successful in resolving the discomfort reported by the evaluee. Extraneous information identifying the employer, their address, and the evaluator are excluded. For that reason they are termed "excerpts," although the essential information regarding all ergonomic aspects remains intact. (A complete evaluation example appears at the end of Chapter 3 and is included in this chapter as excerpt 8.)

Each excerpt is titled with the body part affected so that readers can readily access the excerpts that most closely resemble situations being dealt with in their own workplace. The situations portrayed are the most common circumstances culled from more than 4000 evaluations conducted in typical office environments. You will see some duplication in the areas of discomfort and workstation configuration, but the subtle differences, and combinations of discomfort, will allow readers to find particular cases that most closely compare to situations they face.

A fictitious employee name is assigned to each case to identify the cases with a more familiar reference tag, thereby making it easier to refer back to a case. The most pertinent photos from the original evaluations are included within the excerpts and have been resized for the page layout of this book, which is smaller than the original reports. The

Safety Manager's Guide to Office Ergonomics, By Craig Chasen
Copyright © 2009 John Wiley & Sons, Inc.

employer for every situation is identified as Acme Corporation in place of the actual employer.

1. ROGER SHACKLEFORTH: *ELBOW*

History

Roger Shackleforth is right-handed, 6 ft 4 in. tall, and has worked at Acme for three years. He has been in his current office for one year and has recently been diagnosed with right elbow tendonitis. He states that his symptoms began about six months ago, and during a recent three-day vacation his symptoms diminished but returned quickly when he came back to work. Mr. Shackleforth wears contact lenses during his workday, works 45 to 50 hours a week, and spends 80% of his workday at his computer.

Job Activities

Sitting
Computer input activities: mousing and keyboarding
Minimal handwriting and phone work

Observations

Mr. Shackleforth works at a 29-in.-tall desk with an angled keyboard and wheel mouse on the desk surface. He has two monitors elevated on risers, and his horizontal line of sight to his main monitor meets the screen about 1.5 in. down from the top. He demonstrates a neutral head and neck posture as he views his monitor.

Although he is right-hand dominant, he has been operating his mouse with his left hand since October, in an effort to mitigate his right elbow discomfort. He has been using an angled keyboard for a long time, and two weeks ago he elevated the front edge of the keyboard on the recommendation of his physical therapist. The resulting posture helps decrease his wrist extension, which is the goal of this change. This modification is partially beneficial when he is typing; however, as he views his secondary monitor, to the left of his primary monitor, his input activities are primarily mousing activities. As he is using his mouse with his left hand, and not typing, he demonstrates a tendency to rest his right wrist on the desk in front of the right side of the keyboard. This places his right wrist in significant extension (approximately 45°), which is counterproductive.

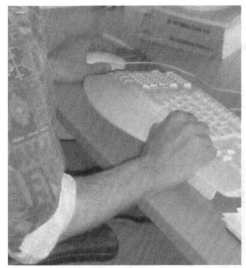 

Initial positioning with wrists above elbows.   Sitting on phone books to elevate elbows.

Mr. Shackleforth sits in a semiadjustable task chair that has very worn cushioning and does not provide adequate support to his upper back or upper legs (toward the knee) because of its small size. The initial height of the seat pan was 19 in. from the floor, which Mr. Shackleforth stated was the highest that the chair will go. As the height was reviewed because his hips were positioned slightly lower than his knee height, it was found that the chair could rise another inch, to 20 in. from the floor. This brought his hips close to knee height, and he reported that the chair cylinder loses its height setting over time and that he does not readily realize that he is slowly sitting lower.

Because of the low seat pan height compared to the 29-in.-high desk and his 6 ft 4 in. stature, his wrists are significantly higher than his elbow height when he types. This posture is not recommended because it often causes a person to flex his or her shoulders while typing and to lean forward away from the chair's backrest.

The armrests are not adjustable and are at a fixed height of 26 in. when the chair is fully elevated. Mr. Shackleforth's seated elbow height is 25 in., which requires him to elevate his shoulders to rest his forearms and elbows on the armrests as he types.

While observing his gait as he walks ahead, he demonstrates that his right shoulder is significantly higher than his left shoulder, which may contribute to, or be the consequence of, his working posture.

As he uses his phone, he demonstrates a posture of holding his phone up to his ear with his right hand raised and his right elbow in a very tight bend. This posture is a likely contributor to his elbow discomfort.

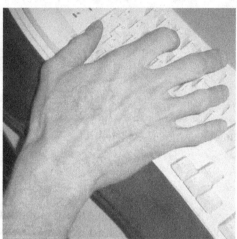

Phone posture: bent elbow and forward head.   Ulnar deviation for backspace key.

Objectives

The primary objectives for Mr. Shackleforth are (1) to allow him to work with his wrists higher than his seated elbow height, (2) to allow him to sit higher with more upper back support and have his shoulders in neutral postures, and (3) to position his keyboard so as to reduce his wrist extension and ulnar deviation.

Recommendation 1

Use a chair with a seat pan that can rise up to 23 in.

Outcome Expected   To position his wrists below his seated elbow height, he was asked to sit on two phone books. This elevated his popliteal height to approximately 23 in. and brought his wrists below his seated elbow height. Therefore, the chair that he uses should have a seat pan that can rise to 23 in. from the floor. This seated height will also provide a better lower extremity posture by raising his hips above knee height.

Recommendation 2

Use a chair with a tall backrest, like the others in the area.

Outcome Expected At the time of the evaluation, Mr. Shackleforth was allowed to try a co-worker's chair with a backrest that is much higher than the one he is currently using. The backrest of his existing chair (shown below) was raised to its highest position, but it is too small to provide adequate back support. The existence of support to the upper back (as well as to the lower back) invites the user to sit back and maintain a more neutral posture along the entire spine. In his current task chair, Mr. Shackleforth demonstrates a tendency to arch his upper back forward, which is contributed to by his wrists being higher than his elbows as he types.

  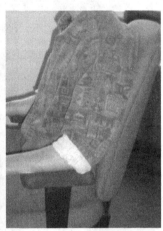

Current backrest set at lowest position.   Current backrest set at highest position.   Chair with tall backrest.

Recommendation 3

Position the keyboard so that his wrists are neutral.

Outcome Expected Mr. Shackleforth's version of the angled keyboard has a forward elevating support that raises the front of the keyboard. This positions his elbows and forearms in neutral postures after his chair is elevated but he still has some wrist extension. Since his desk cannot be lowered and there will not be room for a keyboard tray (especially one with a sharp negative tilt), he may wish to elevate the keyboard in another fashion. The evaluator has often used common doorstops for this purpose because they grip well and the height can be modified easily through their positioning. One small doorstop can be placed under each end of the keyboard at the front edge. During the evaluation, a small stack of note pads was inserted under the front support.

Recommendation 4

Use a gel wrist rest that is designed for an angled keyboard.

Outcome Expected   An available gel wrist rest that is designed for the style of keyboard he uses was added to the plastic wrist support to provide cushioning and to further decrease the angle of his wrist extension.

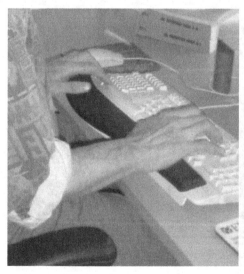

Inclusion of wrist rest for cushioning and angle.

Raising the front more with note pads.

Recommendation 5

Consider a hands-free headset for the phone or the speaker.

Outcome Expected   To eliminate the tight bend in his right elbow during phone conversations, he is encouraged to use a hands-free headset for his phone or the internal speaker on his phone so that he does not need to hold the receiver up to his ear.

2. JANET TYLER: *ELBOW, FOREARM*

History

Janet Tyler is right-hand dominant, 5 ft 10 in. tall, and has been employed at Acme for 14 years, presently as an information associate. She has a diagnosis of right forearm pain with right ulnar nerve symptoms and has been at her current computer workstation for 11 months. Ms. Tyler wears bifocal glasses during her workday. Currently, she works 5 days/ week for 8 hours/day.

Ms. Tyler estimates that she spends 95% of her day at a computer with minimal phone activities. She does very little handwriting and works regularly from reference documents for computer input.

According to Ms. Tyler, last fall she began experiencing soreness in her right forearm, which came and went without significant discomfort. She reports that since the beginning of the year, the symptoms of soreness have not gone away and that she has been experiencing a sensation of tingling her right forearm and wrist. She also reports numbness in digits 4 and 5 of her right hand. She is receiving physical therapy and has work restrictions that require a 5- to 10-minute break every hour.

Job Activities

Repetitive/static upper-extremity motions
Sitting stationary
Static wrist postures

Observations

Ms. Tyler works at a modular corner workstation where the work surfaces are set 29 in. from the floor. The workstation has a cutout section where an articulating keyboard tray allows her to position her keyboard and mouse in a variable range. She reports that her computer was upgraded at the beginning of the year, approximately the time that she began experiencing right forearm discomfort.

Ms. Tyler states that her new computer monitor is smaller and the keyboard is wider. (This keyboard is 18 in. wide and her previous keyboard was 16 in. wide.) She also received a new keyboard tray that is wide enough for the keyboard and mouse. Her previous keyboard tray was not wide enough to accommodate her mouse, and she operated it on the desk surface. She still works with her mouse on the desk because as she sits in front of the home keys of the new, wider keyboard, the mouse is positioned far to the right and requires her to reach outward.

Her keyboard tray is positioned at a 20° positive angle, which incurs significant wrist extension. Her initial wrist angles were measured as follows:

Right wrist extension for keyboard	45°
Right ulnar deviation for keyboard	25°
Left wrist extension for keyboard	50°
Left ulnar deviation for keyboard	10°
Right wrist extension for mouse on desk	35°
Right wrist extension for mouse on tray	45°
Right ulnar deviation for mouse	15°

She is able to press her right thumb against her right volar forearm (with a left-hand assist), demonstrating hypermobility in her wrists (especially right) as she inputs on her keyboard.

Extending right arm to mouse on the desk.   Tendency to ulnar deviate her right wrist.

Recommendation 1

Adjust the keyboard platform to a horizontal position.

Outcome Expected   At the time of the evaluation, Ms. Tyler was shown how to adjust the angle and the height of her keyboard tray. It was moved into a more horizontal position so that it reduced her wrist extension and did not hit her legs, which had impeded her ability to get close to the keyboard. This modification allowed her to get in closer, with her right upper arm parallel to her torso as she types. This more horizontal position decreased her wrist extension by approximately 30%. The new measurements were:

Right wrist extension for keyboard	30°	15° of improvement
Left wrist extension for keyboard	35°	15° of improvement
Right wrist extension for mouse on tray	25°	20° of improvement

Recommendation 2

Try a keyboard that is not as wide and has an angled key layout.

Outcome Expected   To decrease her ulnar deviation and reduce her right arm abduction to access her mouse on the keyboard tray, a smaller keyboard was installed. The small keyboard is 15 in. wide, which is 3 in. narrower than her current keyboard. (It does not have a numerical keypad, but the keys are full sized.) This keyboard can be adjusted through a varying range of angles to reduce ulnar deviation. The evaluator will leave this keyboard for one week to determine if it provides relief of her symptoms. Other keyboards are available that provide the same features.

Split keyboard to keep right wrist neutral.   Keyboard tray raised with upper arm neutral.

Recommendation 3

Raise the chair back to the highest position.

Outcome Expected   At the time of the evaluation, the backrest of her chair was raised to its highest position, which is approximately 3 in. higher than it had been. This provided increased upper back support, and she noted immediate comfort from this adjustment. This will induce her to sit back more readily and thereby reduce upper torso muscle recruitment.

Recommendation 4

Raise the monitor.

Outcome Expected   Ms. Tyler demonstrates that she views her monitor through the upper lens of her bifocals, especially since the new smaller monitor can be moved back farther (27 in. from her eyes). Accordingly, she flexes her cervical spine slightly as she views her new, smaller monitor placed on the work surface. Her monitor was placed onto a 1.5-in.-thick book to raise it slightly. She will try using this monitor height and report any effects to her physical therapists and her physician.

Recommendation 5

Try using a trackball in place of a mouse.

Outcome Expected   Because Ms. Tyler feels significant discomfort from mouse activities, she may benefit from an alternative mousing device. She was shown a trackball that is being used by a co-worker who had symptoms similar to those of Ms. Tyler. That trackball is manipulated primarily by thumb activities, and many other trackballs engage finger manipulations to direct the cursor movement. It may be beneficial for her to try a trackball as a mousing device to reduce arm movements.

3. LEW BOOKMAN: *ELBOW, FOREARM, NECK*

History

Lew Bookman is right-hand dominant, 6 ft 3 in. tall, and has worked as an international audit supervisor at Acme for 21 months. He has been in his current cube for three months and estimates that he works 45 hours a week, with 75% of his time spent at his computer. He wears reading glasses at all times and his prescription was recently modified by approximately 10%.

He reports that he experiences discomfort, tingling, and muscle knots in his upper back and right arm. He states that for more than a year he has felt knots in the center of his upper back at the base of his neck, as well as in the inside of his right elbow. He also reports tightness in the lateral and medial aspects of his right forearm and feels tingling in his right hand. He notes that these symptoms tend to subside during time away from work and are less pronounced at the beginning of the workday than at the end. He reports that extensive periods of mouse (trackball) activity will aggravate his symptoms. He has been receiving weekly therapeutic massage treatments for many months but his discomfort remains at a plateau.

Job Activities

Sitting stationary
Static wrist and arm postures
Repetitive upper-extremity motions

Observations

Mr. Bookman works at a modular corner workstation that has the work surfaces set 28.5 in. from the floor. He uses an angled keyboard and a thumb-motion trackball placed on the work surface. There is an articulating keyboard tray attached under his desk, but the platform is not wide enough for his keyboard and the trackball, so he does not use it. With it retracted under his desk, it makes contact with the top sections of his upper legs, impeding his ability to move his legs under his desk and restricting his leg movement.

His monitor is placed on a 9-in. monitor riser with legs that are stackable cylinders. This height places his horizontal line of sight at midscreen level. He is observed to tilt his head back, extending his cervical spine, to view the upper sections of his screen.

He sits in a fully adjustable task chair that he has raised to its highest position: 22 in. from the floor. At this height, his hips are at the same height as his knees, and the seat pan is not deep enough to provide sufficient support to his upper legs. The chair's backrest is small and will adjust to his upper or lower back, but it is not tall enough to provide support to both. He has it positioned at the lowest setting so that it provides support to his low back. He is observed to slide his hips forward occasionally so that his back moves down on the backrest, allowing his shoulder blades to be supported by the backrest. This provides temporary relief to his upper back but allows his lumbar spine to "round" without support, since his hips are moved forward. His chair has an adjustable-depth seat pan currently set in the middle position.

He has approximately one hour a day of phone work and demonstrates that he must often cradle his phone between his shoulder and neck as he works on his computer. This is a significant stressor to his upper extremities.

His cube has full windows along the wall behind him, which allows a large amount of natural light into his work area. This creates glare on his monitor screen, especially in the afternoon. As he was being reviewed, he was observed to move his head in short deliberate movements while he viewed different sections of his screen. As these movements were tracked, it was clear that he places his head in the path of the light through the windows onto the screen area he is viewing. He does this to shade the glare from the direct light on the screen area that he is viewing. This causes him to engage his upper torso muscles in constant tension to move and position this shadow from his head onto the screen area he is viewing. This brings his shoulders forward and places his right elbow at an angle of approximately 75° as he operates the mouse. This adds pressure on his right forearm against the desk and creates an awkward shoulder motion when it is forward of his elbow. There is also a 2 ft × 4 ft recessed fluorescent light fixture directly above his monitor, which adds to the glare.

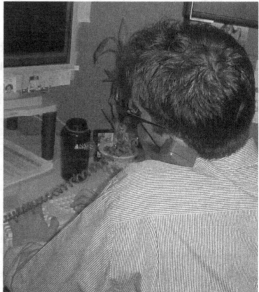

Forward lean as he blocks natural light glare.

Cradling phone as he types on his computer.

Mr. Bookman does not use a wrist rest for his trackball, calculator, or numerical pad on his keyboard. Because of this, he rests his forearm against the work surface, which causes contact pressure that can be damaging, as the tissues are compressed over time.

Wrist bent back with contact pressure against desk.

A wrist rest reduces the angle and cushions the contact.

Recommendation 1

Remove the unused keyboard tray from under the desk.

Outcome Expected To allow more legroom under his desk, the unused keyboard tray should be removed.

Recommendation 2

Consider moving to a cube away from windows and overhead lighting.

Outcome Expected To reduce his head and neck postural changes caused as he blocks the light through the windows and the light fixture directly overhead, he is advised to try working in a cube without such light sources. Mr. Bookman advised that a nearby cube, along the hall leading to his area, is vacant and is away from windows. Additionally, the fluorescent light fixture in that cube is not directly above the computer station as it is in his current cube.

Recommendation 3

Use keystroke alternatives to reduce trackball activities.

Outcome Expected   To reduce his right arm activity with his trackball, he is advised to learn and use the many keystroke alternatives that eliminate the need to use mouse clicks. An online list of keystroke alternatives to mouse clicks is available at the evaluator's website.

Recommendation 4

If possible, obtain a chair with a full backrest and deeper seat pan.

Outcome Expected   Because of his height, his current chair does not provide adequate dimensions for seat pan depth and backrest height. His current chair has an adjustment for the seat pan depth, which was positioned at middepth. It was moved forward to the longest depth but still does not provide adequate upper leg support. He would benefit from a chair with larger dimensions in those aspects. Additionally, having a backrest that provides support for his upper and lower back would be beneficial. Secondarily, a larger chair would probably rise higher, allowing him to place his hips above his knees and sit with a more neutral posture in his lower extremities.

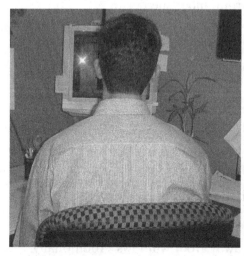 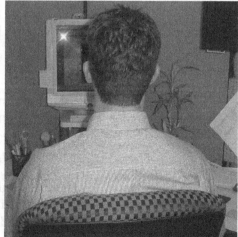

Backrest at the initial, lowest, position.   Backrest at the final, highest, position.

Recommendation 5

Raise the chair armrests for forearm support.

Outcome Expected   To allow him to get support to his arms without resting them on his work surface, his chair armrests were elevated so that they are positioned just below his forearms. This was beneficial as long as they are not elevated to a height that causes him to flex his shoulders.

Recommendation 6

Lower the monitor to reduce his neck flexion to view the upper sections of the screen.

Outcome Expected   At the time of the evaluation, one 1-in. section of the stacking legs on the monitor riser was removed, which lowered it from 9 in. to 8 in. above the desk surface. He reported that he felt immediate improvement from this modification.

Recommendation 7

Work with wrist rests for the trackball, keyboard, and calculator.

Outcome Expected   To reduce his wrist and forearm contact pressure and place his wrists in a more neutral posture, he was allowed to try gel wrist rests. These were placed at the trackball and keyboard number pad as a trial. These were very helpful in cushioning his wrists as well as placing his wrists and shoulders in more neutral postures. The items tried were a special gel that is malleable enough to avoid pressure points and to distribute the arm weight evenly. A palm rest designed for an angled keyboard is readily available from office supply companies.

Recommendation 8

Perform "desk stretches" during microbreaks.

Outcome Expected   To promote blood flow and allow muscle recovery during his workday, he is advised to stretch at his desk during very short microbreaks. The stretches were described in a separate document that he should have reviewed by his therapist to ensure that they are consistent with his treatment plan.

4. ARCH HAMMER: *ELBOW, SHOULDER, NECK*

History

Arch Hammer is left-hand dominant, 5 ft 10 in. tall, and has been a software engineer at Acme for three years. He has worked at his current workstation for approximately eight months. He reports that he has been experiencing pain that begins at the top of his left shoulder and migrates down the outside of his left upper arm into his elbow. He also notes discomfort at the back of his neck where it meets his shoulders.

He states that his symptoms began from a cabling project during which he was assembling 100-ft rolls of fiber-optic cable. He reports that one day about four months prior to the evaluation, when he was carrying four rolls over his left shoulder, he felt the initial onset of shoulder discomfort. He states that his symptoms have not diminished and are continual. He has been seen by his personal physician and will begin physical therapy treatments in two days.

Observations

Mr. Hammer works 40+ hours a week at a modular computer workstation with his work surfaces positioned 28.5 in. from the floor. He has an articulating keyboard tray at the curved corner surface and had been working with a mouse placed on the left side of the tray. He reports that one week ago he moved the mouse to his right (nondominant) side and has been able to perform his work with reasonable accuracy. This is especially beneficial because he uses a program that requires a significant amount of "cutting and pasting." To perform those functions, he uses his left hand to press the "Ctrl" and "C" or "V" keys simultaneously. When pressing those keys with his left hand and also using his left hand for the mouse, he would be moving his left hand and wrist repeatedly between the keyboard and the mouse (with his wrist in an awkward posture).

Mr. Hammer sits in a semiadjustable chair with the tilt tension of the chair's backrest at its lowest setting. As a result of having very limited resistance from the chair back, he sits in a very reclined posture. His 21-in. monitor is placed directly on the work surface and when sitting in a reclined position, his neck (cervical spine) is flexed forward signifi-

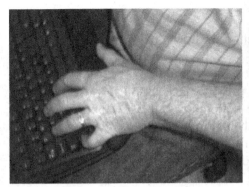
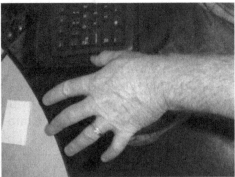

Pressing "Ctrl" and "C" or "V" with left wrist in ulnar deviation, close to his torso.

Left hand in previous position for mousing on the left; moderate ulnar deviation in the left wrist.

cantly. He also works from documents which he places on the work surface in front of his monitor, where viewing them also requires him to bend his neck forward.

His chair does not have a control that allows him to lock the chair back into an upright position. (The chair instructions show that function as an option for his chair, but it was not included.)

He has positioned his articulating keyboard tray with the front edge 31 in. from the floor and the rear edge 32 in. from the floor. This positive angle requires him to place his wrists above his elbows, and he incurs moderate wrist extension (bending back) as he types as well as ulnar deviation.

Initial posture: keyboard at 32 in. from floor and chair reclined.

Final posture: negative tilt keyboard and back supported.

Recommendation 1

Tighten the tension control of the chair and try a back cushion.

Outcome Expected   At the time of the evaluation, the tilt tension control of his chair was tightened to provide back support. It was tightened from its lowest setting to the maximum, but Mr. Hammer still reclined partially as he leaned into the chair backrest. A wedge-shaped cushion was tried as a way to keep him in a more upright position. The cushion is 3 in. thick at the wide wedge dimension and allowed him to sit in a much more vertical posture. He stated that this was very comfortable. The back cushion is available from local suppliers, although another chair that has a control to lock the tilt angle would provide the same benefit more easily.

Recommendation 2

Raise the monitor.

Outcome Expected   At the time of the evaluation, a 3-in.-thick phone book was placed under his monitor to help reduce his neck flexion. This was successful, although his eyes still meet the screen at the very top of the monitor. Ideally, the monitor should be raised approximately 2 to 3 in. higher in a few days so that he can adjust to this neck posture change incrementally.

Recommendation 3

Try raising reference documents between the monitor and the keyboard.

Outcome Expected   At the time of the evaluation, two large three-ring binders were placed between his keyboard and his monitor (now raised 3 in.) to reduce how much he must bend his neck down to view the documents. This was beneficial, but the documents should be raised higher, especially if the monitor will be raised further. A document holder designed for this purpose and an adjustable-height monitor riser are both available from the office supply vendor.

Recommendation 4

Position the keyboard lower on a negative angle to reduce wrist extension.

Outcome Expected At the time of the evaluation, his keyboard tray was adjusted 2 to 3 in. lower and placed at a slight negative tilt. This places his wrists below his seated elbow height and reduces his wrist extension as he types.

Recommendation 5

Use a small gel wrist rest to reduce wrist extension for mousing.

Outcome Expected At the time of the evaluation a short gel wrist rest was placed in front of his mouse to provide cushioned support for his wrist and palm to reduce his wrist extension. This will be beneficial for either hand that he uses for his mouse, and he should ensure that he purchases a soft gel product where he rests the thenar pads of his hands more than at the wrist crease.

5. JEREMY WICKWIRE: *FOREARM*

History

Jeremy Wickwire is left-handed, 5 ft 9 in. tall, and has been an account executive at Acme for 10 years. He has been experiencing significant discomfort in his forearms L > R, with less discomfort in his shoulders. He reports that approximately six months prior to the evaluation, he noticed discomfort along the top of his left forearm. Initially he attributed the discomfort to weight lifting; however, it has not diminished and he has not been lifting weights since that time.

Observations

Mr. Wickwire works in a small room with a 29-in.-tall pedestal desk that is 18.5 in. deep. He has his laptop installed in a port replicator on the desk surface and uses an external mouse with his right hand. He sits in a semiadjustable conference room–style chair that has adjustments for height and tilt tension only. He has the height set at 22 in. from the floor, with the tilt tension at its lowest setting.

Sitting at this height, his feet do not rest on the floor unless he is sitting forward on the seat pan. This posture is also caused by the tilt tension of the backrest being so low that if he leans back, the backrest reclines rearward with no resistance or support.

The port-replicator elevates the laptop so that his wrist rest is nonfunctional and places his wrists higher than his elbows as he types. His horizontal line of sight is approximately 5 in. above the top of the laptop monitor, and he must flex his neck to view the screen.

With laptop on desk, his wrists are higher than his elbows, arms are extended, shoulders are rounded.

With keyboard and mouse at seated elbow height, he can sit and receive support from his chair-neutral posture.

When Mr. Wickwire travels, he carries his laptop in a carrying case that has compartments for documents and accessories. At the time of the evaluation, the bag weighed 12 lb without the laptop or additional resources. His laptop weighed just under 10 lb, and the total weight of the bag, when Mr. Wickwire is traveling, is estimated at 25 lb. Because Mr. Wickwire is left-hand dominant, he demonstrates a natural inclination to lift and carry the bag with his left hand and arm.

Mr. Wickwire wears his watch on his left hand and had been wearing a large metal-banded watch that is unusually heavy for a wristwatch. He reports that he no longer wears this watch.

Objectives

The primary objectives for Mr. Wickwire are (1) to allow him to sit with his chair lower, (2) to have the backrest provide more back support, (3) to position his keyboard and mouse at his seated elbow height, and (4) to raise his monitor.

Recommendation 1

Lower the chair and readjust the seat pan and backrest.

Outcome Expected At the time of the evaluation, Mr. Wickwire was asked to lower his chair so that his feet reached the floor, with his hips slightly higher than his knees. The resulting height for neutral lower extremity postures was 3 in. lower: 19 in. from the floor.

Recommendation 2

Tighten the tilt tension on the chair.

Outcome Expected To have his conference room–style chair provide more upper back support, the tilt tension was increased from its lowest setting to its highest setting (100%). This provided significantly more back support and allowed him to lean back against it as he works.

Recommendation 3

Use an external keyboard and mouse on an articulating tray.

Outcome Expected After moving his chair lower, his shoulders and arms are elevated even more dramatically as he operates the keyboard and mouse of his laptop computer. His desk is equipped with a pull-out keyboard tray 26 in. wide positioned 25¾ in. from the floor. The evaluator provided a keyboard for him to try on the tray, which placed his wrists lower than his elbows, which is optimal. The keyboard used did not have an embedded numerical keypad on the right side, which makes it less wide and provides more room for a mouse, since the tray is only 26 in. wide.

Recommendation 4

Elevate the laptop so the screen is higher.

Outcome Expected   Although he is now sitting 3 in. lower, the laptop screen is still quite low and causes him to tilt his head down to view the lower sections of the screen. Because he will be using an external keyboard and mouse, he can raise the entire laptop to elevate the screen. He is advised to raise it so that his horizontal line of sight meets the screen one-third of the way down from the top.

Recommendation 5

Try to reduce the weight of items placed in the travel bag.

Outcome Expected   To reduce the weight of his laptop travel bag, he is advised to redistribute some of the contents into other bags. Also, there are a variety of lighter laptop carrying bags, as well as lighter laptop computers.

6. LIZ POWELL: *FOREARM*

History

Liz Powell is left-hand dominant, 5 ft 5 in. tall, and has been employed as a Q/A monitor at Acme for 12 years. She has worked at her current workstation for one month and since then has been experiencing discomfort in the underside of her forearms near her wrist crease (R > L).

Job Activities

Sitting stationary
Static arm and shoulder postures
Finger manipulations

Observations

Ms. Powell works at a corner workstation that is 29 in. tall. She has her keyboard on an articulating keyboard tray that is 22 in. wide, which is only wide enough for her keyboard. Her mouse is placed on the work surface behind her keyboard and approximately 2.5 in. higher, causing her to extend her right arm fully as she uses the mouse.

She estimates that she spends 90% of her day working at her computer, and she does not use wrist rests. Although left-handed, she uses her right hand to operate the mouse and positions her phone on her right but picks up the receiver with her left hand.

Her keyboard is adjusted to a slight positive angle with the front edge 26 in. from the floor. In the place of a cushioned wrist rest, there is a hard plastic lip on which Ms. Powell places the palms of her hands as she types. This places her wrists in a posture of approximately 35° to 40° of wrist extension for both hands.

Fully extends arm for mouse and leans away from chair.

Mouse on keyboard tray with arm back using backrest.

Her monitor is placed directly on the work surface, and the top edge of the screen is just below her horizontal line of sight. When viewing the screen, especially the menus on the bottom, she tilts her head downward approximately 15° from vertical.

She often works from documents, which she places on the desk surface in front of the monitor. Reading documents on the work surface also causes her to tilt her head downward.

Her fully adjustable chair is positioned very well for her lower extremities and back support. However, when using the mouse or looking at her monitor or documents, she will lean forward and move her upper back away from the chair back.

Recommendation 1

Replace the platform of her keyboard tray with a wider one, or use a small keyboard so that she can have the mouse at the same level as the keyboard.

Outcome Expected   To eliminate her right arm reach for the mouse, which also pulls her away from the chair backrest, the mouse should be positioned adjacent to the keyboard. There is insufficient room on the existing keyboard tray, so the 22-in.-wide platform should be replaced with a 25- to 27-in.-wide platform. The existing mechanism can remain in place as long as the wider platform will attach to the existing mechanism.

A new, wider platform would also come with a wrist rest and would eliminate the angle in her wrists for keyboard and mouse work. A gel wrist rest would also eliminate the contact pressure she has on the underside of her wrists from resting them on the hard surfaces she has now. At the time of the evaluation, her keyboard tray was adjusted with a slight negative tilt which eliminated some of her wrist extension when typing. The front edge of her keyboard tray ended up 27 in. from the floor, 1 in. higher that it was.

There are also a variety of smaller keyboards, between 13 and 15 in. wide, which would be small enough to allow space on her existing key-board tray. However, the hard plastic edge on the front of her keyboard platform does not serve as a good wrist or palm rest for mouse activities.

Recommendation 2

Raise the monitor for a neutral neck posture.

Outcome Expected   To prevent her from tilting her head down to view the middle and lower sections of the screen, her monitor should be elevated approximately 3 to 4 in. At the time of the evaluation, two phone books were placed under it, which raised it 3.5 in. and eliminated the flexion in her cervical spine as she views the screen.

Recommendation 3

Use a document holder between the keyboard and the monitor.

Outcome Expected   To reduce her need to lean forward and bend her neck to view documents, she should use a document holder that places documents between the keyboard and the monitor. Initially, she tried a three-ring binder to get a sense of that positioning, which was benefi-cial, so a demo model of such a document holder (shown below) was tried at the end of the evaluation.

Monitor raised and binder under documents.　Monitor raised; document holder designed for that space.

7. RUTH MILLER: *FOREARM*

History

Ruth Miller is right-hand dominant, 5 ft 5 in. tall, and has been employed as an operational support technician at Acme for 12 years. She has a diagnosis of left forearm pain and tendonitis, and has been at her current workstation for nine years. Her job responsibilities are to work with customers who come to the Acme utility offices and to communicate information to the divisions of transportation, utilities, and signs and signals. She works 5 days/week for 8 hours/day.

According to Ms. Miller, she began experiencing diminished strength in her left hand about six months ago. This weakness in her hand was accompanied by dull pain in her mid-forearm, along the lateral side of her left elbow and upper arm and into her upper left scapula. She reports that she has experienced left elbow discomfort for the last eight years as the result of hitting the back of her elbow against the edge of the left-side work surface of her current workstation. She was treated by her personal physician at that time and received physical therapy.

Job Activities

Sitting stationary
Static wrist postures

Frequent extended reaching

Frequent twisting while seated

Observations

Ms. Miller works at a hexagonal counter-workstation with work surfaces 30.5 in. from the floor and counter surfaces at the height of the top of her head. These counters are designed to be at chest height for visitors and co-workers who stand as they work with Ms. Miller.

Ms. Miller has five primary tools:

Computer

Multiline phone

Radio base station

Calendar

Phone record log

These items are positioned in a 180° arc, with the center at the point where visitors approach the counter. On her side of the counter there are shelves to store small items. To accommodate a full-sized computer monitor, a section of the shelving was removed at the right corner of her workstation. She now uses a flat-screen monitor that is 14 in. wide, but she still has it positioned in the far-right corner of her workstation, directly opposite the position of the radio base station. She is observed to control the radio with her left hand while operating the mouse with her right hand positioned near the monitor. This is a very awkward and taxing posture and motion.

There is 36 in. between her radio and her telephone, and 40 in. between her phone and her monitor. She uses her work surface to work with a variety of documents, and to allow space for them, she will push all tools to the back of the work surface. This creates an extended reach for those items, with her phone receiver placed 24 in. from the front edge of the desk. She regularly leans her left elbow and forearm on the work surface and demonstrates wrist extension as she manipulates items such as her calendar (see photo 2). Even when she works on her computer keyboard, she rests her left elbow against the work surface (see photo 5). Her keyboard is positioned 16 in. back from the desk edge with the feet extended, causing moderate wrist extension (see photo 6).

Photo 1: Initial reach for radio base station.　Photo 2: Initial reach for calendar; cradles phone.

Ms. Miller sits in a fully adjustable chair with the seat pan positioned 19 in. from the floor. This height provides an optimal lower extremity posture; but in relation to her 30.5-in. work surfaces, her wrists are slightly above her seated elbow height, which is not recommended.

Her workload on the day of the evaluation was light, so the frequency of her left arm activities was minimal. Independent of frequency, the arm motions and reaching to perform her work are continual, with constant extended arm reaching.

She must often conduct phone conversations while taking notes and performing computer input. To perform these simultaneously she will cradle her phone in her neck and left shoulder (see photo 2).

Photo 3: Phone left (diamond-shaped sign is a location landmark).　Photo 4: Phone moved right; monitor at center.

Recommendation 1

Use a hands-free headset for phone work.

Outcome Expected   At the time of the evaluation, the evaluator presented a catalog of numerous headsets that are very efficient for the work she performs. Ms. Miller had the same catalog and stated that she was in the process of placing an order for a headset.

Recommendation 2

Install an articulating keyboard tray for the keyboard and mouse.

Outcome Expected   To place her wrists at, or slightly below, her seated elbow height, eliminating her rounded shoulders and contact pressure on her left elbow, her keyboard and mouse should be positioned on an articulating keyboard tray. The tray should retract easily so that she does not have to reach over it when using the desk space that will become available in the area in which her keyboard and mouse are now.

Recommendation 3

Remove another 15-in.-wide section of shelving for monitor space.

Outcome Expected   The shelving under the counter restricts monitor placement, and some right corner sections have already been removed. However, her monitor is still 180° from her left-side tools and causes her to turn her neck and torso as she works at her desk. If the next 15-in. section to the left can be removed, she will be able to place her monitor directly in front of her (see photo 4). This section of shelving is not currently serving a critical need (see photo 6).

Recommendation 5

Move the phone to the right and closer.

Photo 5: Contact pressure on left elbow. Photo 6: Monitor in recess where shelves were removed.

Outcome Expected At the time of the evaluation, her phone was moved 14 in. to the right and inward. This reduced her left hand reach significantly. Even with a hands-free headset, she will have some contact with her desktop phone controls (see photos 3 and 4).

Recommendation 6

Move the calendar to the right.

Outcome Expected At the time of the evaluation, her calendar was moved 14 in. to the right. This reduced her left hand reach significantly (see photos 2 and 8).

Recommendation 7

Move the radio base station to the right.

Outcome Expected At the time of the evaluation, her radio base station was moved 14 in. to the right. This reduced her left hand reach significantly; however, it would be optimal to move it even closer. The cord length restricted more movement during the evaluation, and Ms. Miller states that she can request to have the cords extended to bring the base station closer (see photos 1 and 7).

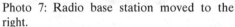

Photo 7: Radio base station moved to the right.

Photo 8: Calendar moved to the right.

Summary

The tools that Ms. Miller uses actively each day are positioned in a wide arc that causes her to engage an extended reach and a torso lean to operate them. Additionally, she pushes them to the back of the 28-in.-deep work surface, which contributes to her extended reaching. Her phone is the most actively used tool, and using a hands-free headset should significantly reduce her left arm motions and awkward postures. Bringing all devices toward the center will also reduce reaching.

Her work surfaces are high relative to her seated posture. The surfaces are at a fixed height of 30.5 in., and installing an articulating keyboard tray would significantly reduce her arm reaching, contact pressure, and rounding of her shoulders for computer work. The desk work surface is 1.5 in. thick and has a 1.5-in. square support tube beneath it. This combined thickness of 3 in. needs to be considered for the installation of a keyboard tray.

8. WANDA DUNN: *FOREARM, NECK*

History

Wanda Dunn is right-handed, 5 ft 4 in. tall, and has worked as a paralegal at her current workstation for one year. For the last two months she has been experiencing tingling in her hands which she reports is especially pronounced when she is editing hardcopy documents. Additionally, she has been experiencing discomfort along the left side of her neck.

Observations

She works at a rectangular desk that is 28 in. tall, with her keyboard and mouse placed on the desk surface and no wrist rest for her keyboard. She has a large "pillow-style" wrist rest for her mouse that she says is very comfortable.

She sits in a small semiadjustable typing chair that places her hips just above her knees, which is optimal, and it has a small backrest at its lowest position. This backrest position does not provide upper back support, and the seat height places her elbows significantly lower than her wrists as she operates her keyboard and mouse. Since her chair does not have armrests, she places some upper extremity weight on her wrists as she rests them against the sharp corner of the desk surface.

Wrist height and no wrist rests create significant contact pressure against corner edge of desk.

Wrists are much higher than her seated elbow height, which is not advised.

Recommendation 1

Install an articulating keyboard tray and move the desk forward, closer to the door.

Outcome Expected To allow her to position her wrists at or slightly below her seated elbow height, she would benefit from placing her keyboard and mouse on an articulating keyboard tray. The width between the side cabinets of her desk is 27.5 in. and there is room to install a full platform (11 in. × 27 in.) keyboard tray. When using the tray, her chair will be moved about 12 in. back from the desk, so she should move her desk forward. There is sufficient space in her office for the desk to be moved forward 1 ft.

Recommendation 2

Replace the small chair with a fully adjustable task chair.

Outcome Expected   During the evaluation, her backrest was raised from its lowest position to its highest position to allow her to receive support to her upper back and allow her muscles to relax as she sits upright. The adjustment that holds the backrest in place will often lose its grip, and she is advised to check it from time to time because the backrest will slowly move down. Because the backrest will not maintain its height and is very small, she is a candidate for a chair with a taller, more adjustable backrest.

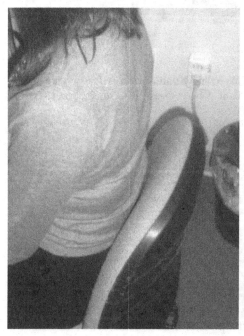 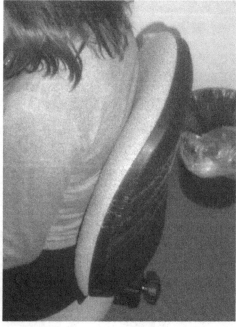

No middle or upper back support from the small backrest.

Even with backrest raised fully there is minimal support.

Recommendation 3

Use a document holder between the keyboard and the monitor.

Outcome Expected With the keyboard tray in place, there will be space on the desk for documents directly in front of her. Rather than placing documents on the flat desk surface and to the side, she should use a wide document holder between the keyboard and the monitor. Most document holders designed for this application will drop the lower edge below desk height so that the documents will not block any part of the screen data.

Looking left and down for hardcopy documents.

Neck remains neutral with documents raised and centered.

9. DORIS RICHARDS: *FOREARM, SHOULDER, NECK*

History

Doris Richards is right-handed, 5 ft 1 in. tall, and has worked as a catalog librarian at Acme for eight-and-a-half years. She has worked at her current workstation for approximately two months, when the new building opened.

She has been experiencing discomfort along the right side of her neck, which migrates down across her right shoulder to the top of her right arm. She reports that she had right wrist discomfort a few years prior but that she became symptom-free after an ergonomic evaluation was performed and the recommendations were implemented. Her current discomfort has arisen since her move to her new office, and she notes that it dissipates during weekends and when she is away from work.

Observations

Ms. Richards works at an L-shaped desk with 28-in.-high tall work surfaces. Her computer station is positioned on the left surface, which is 20 in. deep, where there is a 24-in.-wide articulating keyboard tray set 26 in. from the floor. Ms. Richards has her standard keyboard and mouse placed on the keyboard platform, and the mouse space is quite small left to right.

As she operates the mouse, she is observed to position her right elbow forward, with her right shoulder elevated higher than a neutral posture. Her left shoulder is also observed to be significantly elevated.

She sits in a fully adjustable task chair that she positions 18.5 in. from the floor, which prevents her feet from resting firmly on the floor. She uses this height to raise her upper extremities to the 28-in.-high work surfaces for computer work and for handling numerous books that she catalogs in the course of her work. She has a footrest at her computer station, and she uses an old book as a footrest on the right-side section of her desk. She states that these footrests help her sit comfortably as she works.

At the time of the evaluation, her chair's backrest was positioned so that the lumbar protrusion was aligned slightly below her lumbar lordosis (inward curve), and the adjustable depth was extended to 80%. This placed the lumbar protrusion at the top of her hips and created a gap between the backrest above the protrusion and her middle and upper back.

Her seated elbow height is 25.5 in., yet her chair armrests are elevated to 27.5 in. from the floor. This positioning creates significant shoulder flexion as she operates her keyboard and mouse and is probably why she extends her elbows forward to lower her shoulders slightly as she works. As she views her monitor, her horizontal line of sight meets the monitor approximately one-third of the way down from the top of the screen, which places her head and neck in neutral postures for all sections of the screen.

Keyboard at 26.5 in. and armrests at 27.5 in. causes her to work with her elbows forward and shoulders elevated.

Keyboard down to 25 in. and armrests at 25.5 in. places elbows close and lowers shoulders. Backrest raised 1.5 in..

Recommendation 1

Adjust the backrest to align with her spine, and lower the armrests.

Outcome Expected At the time of the evaluation, her chair's backrest was readjusted so that the lumbar protrusion was raised 1.5 in., allowing it to align with her lumbar lordosis (inward curve). Additionally, the depth of the adjustable lumbar protrusion was reduced from about 80% to 0%, so that her shoulder blades are in contact with the upper section of the backrest, to allow her muscles to relax as she sits upright. Extending the lumbar support to 80% moved her torso forward, reducing her contact with the rest of the chair backrest. Because her seated elbow height when sitting 18.5 in. from the floor is 25.5 in., her armrests were lowered to 25.5 in. so that her shoulders are in a neutral posture as she rests her forearms on them.

Recommendation 2

Lower the keyboard tray so that her wrists are at, or slightly below, elbow height.

Outcome Expected To allow her to work with her wrists at, or slightly below, her seated elbow height for her keyboard and mouse, her current keyboard tray was lowered 1.5 in. This height aligns her elbows with her armrests, allows her upper arms to be close to her torso with her shoulders square, and positions her wrists in a neutral posture.

Recommendation 3

Consider using a smaller keyboard for more mouse room.

Outcome Expected Because most keyboard trays are 27 to 28 in. wide, and Ms. Richards's keyboard tray is 24 in. wide (as all the new trays appear to be at the new library), there is very limited mouse space, with a keyboard that is the typical 19.5-in. width. At the time of the evaluation, she was allowed to try two smaller keyboards that do not have a numerical keypad on the right and thereby allow the mouse to be positioned closer to the center. This creates more mousing room for more relaxed mouse movements and also reduces her right arm abduction outward, which creates a more neutral shoulder posture, in concert with the other changes in shoulder posture.

Keyboard tray high, elbows forward.

Keyboard tray lowered, elbows moved back.

Keyboard with no number pad on right.

Recommendation 4

Try increasing the mouse "speed" to reduce mouse movement.

Outcome Expected At the time of the evaluation, Ms. Richards was shown how to increase the "acceleration speed" of her mouse. The effect this has is to reduce the distance required by the mouse to move the cursor across the screen. The default speed is 50%, and Ms. Richards was quite adept at using this function, even though initially she increased it to only 75%.

Recommendation 5

Try using more keystroke alternatives to reduce mouse use.

Outcome Expected   Since Ms. Richards's use of her mouse requires a long reach from her home keys to the mouse platform, she would benefit from using more keystroke alternatives. She was shown a list on the evaluator's website, which has a very comprehensive list of keystroke alternatives.

10. PETE VAN HORN: *FOREARM, SHOULDER, NECK*

History

Pete Van Horn is right-hand dominant, 6 ft tall, and has worked as a senior telecom analyst at Acme for 16 years. He has been in his current cube for four years and estimates that he works 45 to 50 hours a week, with 95% of his time spent at his computer. He wears bifocal glasses for all activities except computer work, and he wears a special pair of prescription glasses for computer work.

He has been experiencing discomfort and tightness in his right forearm for a few months, which has not diminished. He reports that his symptoms will lessen during the weekend and he has been seen by his personal physician, who recommended a wrist brace, which Mr. Van Horn wears when his symptoms are severe. He has also been experiencing discomfort at the base of his neck and along the top of his shoulders.

Job Activities

Sitting stationary
Static wrist and arm postures
Repetitive upper-extremity motions; reaching with right arm

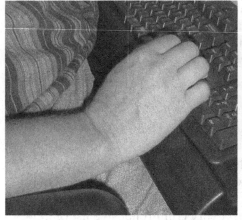 

Mr. Van Horn works at a modular corner workstation that has the work surfaces set 29 in. from the floor. There is an articulating keyboard tray attached under his desk, but the platform is not wide enough for his keyboard and the mouse. Therefore, he has only his keyboard on the tray. He has the tray positioned at 27 in. from the floor and in a slightly positive tilt with recessed palm support. As he types on his keyboard he has approximately 45° of wrist extension (bending back) in both wrists.

Because there is no room for his mouse on the keyboard tray, he places it on the work surface about 4 to 5 in. back from the desk edge. As he operates his mouse in this position, he extends his right arm with an angle of approximately 140° at his right elbow and demonstrates slight wrist extension and contact pressure of his wrist on the work surface. His work is very mouse-intensive, and he is observed to move his right arm frequently between his keyboard and mouse.

Observations

The movement of his right arm between the home keys and the mouse travels through a significant distance and height change, because his mouse is not adjacent to his keyboard. This appears to be the most egregious posture relative to his reported symptoms.

His monitor is placed on a riser that is elevated 3 in. to fit above his laptop computer. This height places his horizontal line of sight just above midscreen level and allows a neutral posture in his neck as he views it. He has his monitor positioned 29 in. from his eyes so that he can place reference documents on the work surface in front of it. Although he has glasses specially designed for computer use and uses them diligently, he is observed to lean forward to view his screen. This forward posture creates significant flexion in his thoracic and cervical spine, and when his right arm is on the keyboard, his right elbow angle is reduced to approximately 70°.

He sits in a semiadjustable task chair that he has raised to a position of 20 in. from the floor. At this height, his hips are slightly higher than his knees. He leaves the tilt function unlocked and is frequently observed to recline as far back as the chair backrest will go.

He does some phone work and is observed to pick up the phone receiver with his left hand. The phone is positioned 16 in. back from the front edge of his desk, on the right side of his monitor, which creates an extended reach for its use.

Recommendation 1

Replace the keyboard tray with a tray that has room for the mouse.

Outcome Expected   To reduce his right arm reach for the mouse, his keyboard tray should be replaced with a tray that is wide enough for the mouse to be positioned to the immediate right of his keyboard. There are such keyboard trays in use in the data center, and these trays typically have a wrist rest for both the keyboard and the mouse, which will be beneficial for him. By moving the mouse off the work surface, there will be room to move the phone forward to reduce the reach distance. Although the keyboard trays in the data center would be an improvement, the separate mouse platform is slightly lower than the keyboard height. The best solution would be for a tray with a 27- to 28-in. platform to be installed so that the keyboard and mouse are at the same height and angle and are adjacent to each other. Often, the platform only can be replaced, and the articulating mechanism is left attached to the desk.

Recommendation 2

Position the keyboard platform in a more horizontal position.

Outcome Expected   To reduce his wrist extension, his keyboard tray should be positioned in a more horizontal orientation. If his tray is replaced with a wider platform, there should be a wrist and palm rest for the mouse to create a more neutral wrist posture as well as reducing significantly the contact pressure against his wrists.

Recommendation 3

Work with a wrist rest for the mouse.

Outcome Expected   To reduce his right wrist extension and contact pressure against the desk, he tried an 8-in. gel wrist rest. It was placed at his elevated mouse position and he reported that it was very beneficial. A similar (but longer) gel wrist rest was left for him to use until his keyboard tray is replaced and/or an 8-in. gel wrist rest can be acquired. Such wrist rests are also available from most office supply vendors.

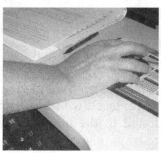

Wrists neutral, keyboard horizontal.   Detrimental contact pressure and angle.   Wrist rest: cushioning and less angle.

Recommendation 4

Try working with the documents closer to him.

Outcome Expected   To reduce his forward leaning and spinal flexion to view his reference documents, two methods of elevating documents were tried. One was to place them initially on an empty three-ring binder, and another was to position them on a document holder designed to position them between the monitor and the keyboard. These tactics allow him to reduce his forward lean, and he will need to balance the reduction of neck discomfort with the potential discomfort from elevating his right arm to write on documents in that elevated position.

Viewing documents on desk.   Documents on three-ring binder.   In-line document holder.

Recommendation 5

Tighten the tilt tension on the chair backrest.

Outcome Expected To provide more back support when sitting upright, the tilt tension was increased from approximately 25% to almost 75% of its maximum. This will allow him to relax his upper torso muscles when sitting upright because the chair backrest is providing more back support.

Recommendation 6

Use keystroke alternatives to reduce mouse activities.

Outcome Expected To reduce his right arm activity for his mouse work, he is advised to learn and use the many keystroke alternatives that eliminate the need to use mouse clicks. An online list of keystroke alternatives to mouse clicks is available at the evaluator's website.

Recommendation 7

Consider operating the mouse with his left hand.

Outcome Expected To allow recovery time for his right arm from mouse motions, he could operate the mouse with his left hand. Like most people who are offered this solution, he is doubtful that he could do it and was reluctant to try this solution. Typically, people are surprised at how easy it is to acclimate to operating a mouse with their nondominant hand, and he may wish to try this when he can. It is also recommended to have a mouse on each side so that precision movements can be performed with the dominant hand, allowing the nondominant hand to perform simple mouse movements.

11. ALLEN REISNER: *LOWER BACK*

History

Allen Reisner is right-handed, 6 ft 4 in. tall, and has worked as a customer services representative at Acme for seven months. He has recently experienced an onset of recurring lower back pain.

He reports that he has maintained employment in the restaurant industry for many years and this is his first job where he must sit during his workday. He also experiences slight left elbow discomfort after his left arm has been positioned in a 90° angle for extended periods.

Job Activities

Sitting
Computer input activities: mousing and keyboarding
Repeated phone conversations
Occasional handwriting

Observations

Mr. Reisner works at a modular corner workstation that has independent crank-adjustable work surfaces for his keyboard and mouse and for his monitor. The front section that holds his keyboard and mouse is positioned 31 in. from the floor, and the monitor section is positioned 39 in. from the floor, which is as high as it will go.

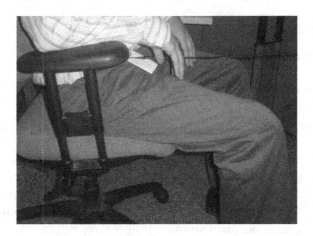

He sits in a semiadjustable chair that he has positioned 22 in. from the floor, which is as high as it goes. This height places his hips slightly above his knees for a relatively neutral posture; however, the seat pan is too shallow for the length of his upper leg. Similarly, the backrest is not tall enough to provide support to his upper back. Because he is unaccustomed to sitting for extended periods, he finds the chair to be very uncomfortable.

He has tried using a lumbar cushion for more lumbar support from his chair, but he states that it is difficult to keep the cushion in place and he is continually repositioning it. Because the cushion is approximately 2.5 in. thick, it pushes him forward in his chair so that the only contact of the backrest with his back is against the area on the cushion, and he then receives even less upper leg support from the seat pan since he is moved about 2 in. forward on the seat pan. He works with his monitor at a distance of 35 in. from his eyes and tilts his head down with moderate neck flexion as he views the screen.

At the time of the evaluation, his work surfaces were raised to their maximum of 39.5 in. to determine if they were high enough for him to work in a standing posture. He felt that the standing posture with surfaces at 39.5 in. was a significant improvement, but it does create an awkward wrist angle since his wrists are well below his elbow height. Two reams of copy paper were placed beneath his keyboard, which raised it 4 in. higher, to 43.5 in. from the floor. This was a much better height for his arms and wrists.

Work surfaces at 39.5 in. for standing and the height-adjustable monitor raised as far as it will go; monitor is still too low.

Keyboard elevated by 4 in. helps wrists and arms attain a neutral posture.

Recommendation 1

Elevate the entire workstation 4 in.

Outcome Expected To allow him to work in a standing posture with his wrists close to his standing elbow height, the facilities department will insert a 4-in.-high frame under the two legs of his adjustable height desk. Because the desk can be lowered to 27 in., he will be able to lower it to a comfortable height for sitting (31 in. from the floor), even with the 4-in. spacers.

Recommendation 2

Elevate the monitor approximately 6 to 8 in.

Outcome Expected   To allow him to maintain a neutral head and neck posture for standing or sitting, the monitor needs to be elevated significantly. He will gain 4 in. from the frame that the facilities department will insert under his desk unit, and he will acquire one or two 2-in.-high "monitor stackers." At the time of the evaluation, two reams of copy paper were inserted under the monitor so that he can get accustomed to the initial (4 in.) elevated monitor height. He will need to reposition the height-adjustable monitor as well as the desk surfaces as he moves between his sitting and standing positions.

12. CHRISTIAN HORN: *LOWER BACK*

History

Christian Horn is right-handed, 5 ft 10 in. tall, and has worked at Acme for five years. He has been in his current cube for about three years and has been experiencing discomfort in his lower back. He reports that he has been diagnosed with degenerative disks at L5/S1 that presented very suddenly about two months prior to the evaluation. His condition arose without incident, but was initially very debilitating. He has since sought medical attention because he has difficulty sitting upright and has requested an ergonomic evaluation to identify any modifications to his work area that will mitigate discomfort as he works.

Observations

Mr. Horn works at a modular corner workstation that has the work surfaces set 29 in. from the floor. He uses a laptop as his main computer with an external keyboard, mouse, and flat-panel monitor, which are all placed on the work surface.

He sits in an old task chair that he acquired from the lab area shortly after his discomfort began. He adjusts the seat pan to 22 in. from the floor for neutral postures in his lower extremities; however, the backrest has poor lumbar support and is unlocked, so he receives no back support unless he is fully reclined back in the chair. This requires that he engage his muscles to sit upright for computer work.

His monitor is positioned so that his horizontal line of sight meets the monitor above the top of the cabinet, which causes him to look down for the lower sections of the screen and pulls him forward away from his chair. He also uses his laptop computer screen at times, which is placed on the right side of his desk, and is very low.

His current semiadjustable chair has nonadjustable back support, is unlocked, and the monitor is low.

Sitting in semiadjustable chair with a small backrest, but it is locked for back support and the monitor was raised.

Recommendation 1

Replace the current chair with a chair that has better back support, along his entire spine.

Outcome Expected At the time of the evaluation, Mr. Horn was asked to try the nonstandard chairs that the EH&S department has made available for employees who have back injuries or for other special requirements. Because of his need for enhanced lumbar support and the occasional need to recline back to approximately 130°, he was allowed to try a co-worker's fully adjustable, tall-back chair with depth-adjustable lumbar support. He found this chair to be extremely beneficial because the backrest remains vertical above the lumbar support, promoting a very neutral spinal posture. This chair is made available because of its exceptional features and adjustability.

Initially, he was provided with a semiadjustable chair with a backrest that is small but elevates through a wide range and has a depth-adjustable seat pan. This chair was an improvement; however, the backrest does not align well with his spine or have lumbar adjustability other than the backrest height.

He was then allowed to try a fully adjustable chair with a tall backrest that stays vertical at the top and has depth adjustment for lumbar support (as well as the backrest height). This chair was very beneficial, the ideal chair for his discomfort issues. It does demonstrate that his seated elbow height is 10 in. above the seat pan height, which is a greater distance than usual. Therefore, he will order one of these chairs with armrests that are taller than supplied typically.

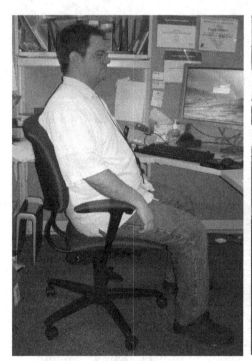 

Semiadjustable chair with a sliding seat pan, height-adjustable backrest, backrest angle adjustment, and armrest height adjustability. This chair is an improvement from his previous chair but does not support his entire spine.

Fully adjustable chair with a sliding seat pan, tall height-adjustable backrest, adjustable lumbar control, and backrest angle adjustment. This chair is optimal for him, and 10-in.-tall armrests will be ordered with it.

Recommendation 2

Raise the main monitor and the laptop monitor.

Outcome Expected The optimal height for monitors, especially to invite users to sit back against the backrests of their chairs, is to have their horizontal line of sight meet the screen one-third of the way down from the top. At the time of the evaluation, his monitor was elevated 2 in. by using "tape cases" from the products with which Mr. Horn works. This gain of 2 in. is approximately half of the height that is recommended, but the monitor height should not be changed more than 2 in. at a time, to prevent the neck discomfort that will usually arise from dramatic changes in monitor height. Because he also uses the screen of his laptop, which is placed on the desk surface, the laptop was placed on a 5-in.-tall riser, to elevate the screen.

13. RICHARD MATHESON: *LOWER BACK*

History

Richard Matheson is right-handed, 5 ft 9 in. tall, and has worked as a client services representative at Acme for one year. He has been in his current work area since starting and has been working with the human resources department to configure his work area in a way to accommodate his lower back discomfort. They recently installed an electric sit/stand work surface for him. He reports that he fractured vertebra L5 in the course of weight lifting and has maintained an aggressive program of stretching, flexibility, and strengthening to manage his symptoms.

Job Activities

Sitting and standing
Computer input activities: mousing and keyboarding
Repeated phone conversations
Handwriting

Observations

Mr. Matheson works at an electric sit/stand workstation that allows him to reposition his computer and phone easily between a sitting and standing position. It is moved electrically and requires no physical force to change the height. The surface has two separate platforms, where the front section for the keyboard and mouse are raised or lowered manually in relation to the main surface that supports the monitor and phone.

He has his semiadjustable chair positioned 21 in. from the floor. He has not adjusted the backrest or armrest height. Both are lower than they should be since they do not align with his spine or reach his elbow height, respectively.

He has the keyboard platform adjusted to 27¾ in. from the floor and in a slight positive tilt. Because of the keyboard angle and lack of a wrist rest, his wrists are positioned in significant wrist extension (bent back 45 to 50°) as he types, which is not recommended.

His horizontal line of sight when sitting or standing meets the monitor about 2 in. above the top of his screen. As a result of this low monitor height, he is observed to slouch and lower his head, which adds curvature to his thoracic spine and rolls his shoulders forward with his head forward of his cervical spine.

When he is standing, Mr. Matheson elevates his work surfaces to the highest position that the main (rear) surface will attain. As he works at this height, he keeps the keyboard surface below the main surface, which causes him to reach down for his keyboard and mouse. This low keyboard and mouse height cause him to round his shoulders and bend his wrists back as he works. The keyboard surface height is 37 in. and his standing elbow height is 43 in., which places his wrists too far below his elbow height.

Keyboard surface below rear surface; tilts head down for low monitor and reaches down for keyboard and mouse.

Keyboard surface at rear surface level; arms/ shoulders neutral, but monitor is still too low.

Recommendation 1

Raise the chair backrest to align with his spine, and raise the armrests.

Outcome Expected   At the time of the evaluation, he was shown how to elevate his chair backrest while seated, and it was raised 2 in. so that the lumbar protrusion aligns with his lumbar curve. His armrests were also raised to provide support without flexing his shoulders.

Recommendation 2

Raise the keyboard surface and place it in a horizontal position.

Outcome Expected   At the time of the evaluation he was shown how to change the angle of the keyboard surface, and it was raised to be 1 in. higher than the monitor surface. This allows more legroom when seated because of the angle, and also reduces the amount of wrist extension when typing in a standing posture since his wrists are lower than his elbows.

When he is in a standing position, the elevated keyboard surface raises his wrists to just below elbow height, which allows him to stand with a more erect spinal posture, with his shoulders and head in neutral postures (with the exception of the monitor position, which is currently too low).

Recommendation 3

Elevate the monitor 4 to 6 in.

Outcome Expected   To help him maintain a neutral spinal posture when sitting and when standing, his monitor should be raised. His horizontal line of sight should meet the monitor about 3 in. down from the top of the screen. Therefore, his monitor should be raised another 4 to 6 in. in addition to the 3 in. of elevation from his current monitor riser.

Recommendation 4

Use gel wrist rests for the keyboard and mouse.

Outcome Expected   To reduce his wrist extension when typing and using his mouse, he should place wrist rests in front of each. Wrist rests with a soft gel cushion are recommended.

14. VICTORIA WEST: *LOWER BACK*

History

Victoria West is right handed, 5 ft 3 in. tall, and has worked at Acme for seven months. She reports that she had an intense onset of lower back discomfort during the first week of January that included an emergency room visit, and she continues with physical therapy and medication in her recovery. She was unable to work until February 4, but has returned to work full time. She has significant discomfort when sitting, and her movements in and out of a chair are performed very carefully so that her lower back discomfort is not aggravated.

Observations

Ms. West works at a new modular workstation with the work surfaces set 29 in. from the floor. She is now sitting on a large Fit Ball, because it supports her more comfortably than a typical chair. She feels that the standard office chair that she has been using is now intolerable. The Fit Ball places her lower extremities in neutral postures and creates a seated elbow height 28 in. from the floor. This elbow height is 1 in. lower than her desk height and creates an awkward posture for her, especially as she uses a mouse. She uses various locations for the mouse and reports that she does not actively use the numerical keypad on her keyboard.

She uses a thin angled keyboard that was selected because of the "quiet keys" and uses a standard optical mouse. She has a flat-panel monitor that is placed on the desk surface, and her horizontal line of sight meets the screen at the very top, which causes her to tilt her head down as she views the lower sections of her screen. She states that she just moved her monitor off a monitor stand (and docking station) that is 5 in. tall. That monitor height felt uncomfortable to her, but she does not report neck discomfort. Still, she cannot achieve neutral postures because her seated elbow height is lower than her desk and keyboard height.

Recommendation 1

Install an articulating keyboard tray that can be positioned above and also moved below the work surface.

Outcome Expected To position her wrists just below her seated elbow height when sitting or standing, she should use a sit/stand articulating keyboard tray. The typical sit/stand keyboard trays rise only 8 in. above the desk surface (37 in.), so her desk surface will be elevated 1 or 2 in. so the tray will rise that much higher since her standing elbow height is 40 in. from the floor. This model tray descends 5 in. below the desk surface on the low end, so she will have her wrists at, or slightly below, her seated elbow height when sitting.

The tray shown above is a tray that has already been used success-fully at Acme and is 30 in. wide, with two arms. The installation width needed is 25 in., and Ms. West is requesting that it be installed at her left-side surface because she does not like to have her back to the entryway of her cube. This will require moving the drawer unit from the left to the right.

Recommendation 2

Move the standard mouse to the left and use a trackball on the right.

Outcome Expected At the time of the evaluation she tried a trackball for her right hand, which reduced the shoulder impacts that mouse movement was causing. She is advised to move her current mouse to the left side for occasional nonprecision work.

Recommendation 3

Replace the standard-size keyboard with a small keyboard.

Outcome Expected At the time of the evaluation she tried a small keyboard that allows her to move her mouse and trackball closer to the center. This model of keyboard does not have an embedded numerical keypad on the right side, so it is 5 in. narrower than her current keyboard. With her 30-in.-wide keyboard platform, she will easily be able to accommodate the small keyboard and a mousing device on each side of it.

Small keyboard allows mouse (right arm) to be close.

Trackball on right and standard mouse on left.

Recommendation 4

Consider using an articulating monitor arm for the sit/stand station.

Outcome Expected Her current monitor stand is not height adjustable, and it would be beneficial if she could easily raise or lower her monitor as she moves between sitting and standing. Because she has a flat-panel monitor, it can be attached to an articulating monitor arm that will easily clamp to her desk surface. (Two-piece clamps are recommended because they eliminate the need to remove the surface to which it is being attached.)

Recommendation 5

Consider using a fully adjustable chair with adjustable lumbar support.

Outcome Expected   Ms. West still has extreme difficulty sitting in any standard chair. During the evaluation she tried most of the chair styles that were available. She tried the newly purchased chair of a co-worker. This chair, when adjusted for her, provided better back support but the "air pump" lumbar support was not included on this model. The air pump can be inflated to the specific dimension that she needs, and the backrest provides optimal upper back support because the top remains relatively vertical and stays supportive to her upper back.

15. SYLVIA BAYLES: *LOWER BACK, LEG*

History

Sylvia Bayles is right-hand dominant, 5 ft 3 in. tall, and has worked at Acme for four years. She has been experiencing discomfort in her lower back and legs since a new chair was introduced for her use. Her department is a monitoring center which is visible to customers, and they had new chairs brought in to create uniformity of appearance.

She reports that she underwent a number of lumbar spine surgeries many years ago, which had been successful and allow her to be relatively pain-free. Her symptoms had been atypical for lower back injuries, where the pain and numbness presented along the tops of her legs rather than the back and sides of the legs, where radiculopathy is typically felt.

Ms. Bayles works Friday through Monday in shifts of 13, 12, 12, and 4 hours a day, respectively. She states that she was issued a "new chair" (the manufacturer's name was not found) at the start of work on a Friday, and that after two hours of use, she felt pain that was similar to the presurgery discomfort she had experienced years before, as well as discomfort in her lower back. She then retrieved an old semiadjustable task chair with a midsized backrest that she had used before the new chair, to finish her shift. She used that chair through the rest of her 4-day workweek and felt no further discomfort.

When she returned to work the following week, she felt that she should try working in the new chair, and while she adjusted it to the best of her ability, it was very uncomfortable and she notes that she feels like she is sliding off the front edge as she sits with the backrest in a vertical position for back support. She states that her symptoms returned rapidly, and that she could not walk with her normal gait for two days after the workweek she spent in the new chair.

Job Activities

Computer and phone work
Sitting stationary
Static wrist and arm postures

Observations

Ms. Bayles works at a height-adjustable computer monitoring station, where her keyboard and mouse are placed on the work surface. Like all work areas in this group, her work area is shared with other employees in a 24/7 operation. She has three flat-panel monitors that are all height-adjustable.

Because her only reported discomfort was triggered by the arrival of the new chair, the new chair and her seated postures were the primary focus of the evaluation. The new chair has four adjustment controls that allow adjustment for the (1) seat pan height, (2) backrest proximity to the seat pan (forward/back), (3) tilt tension of the backrest recline, and (4) angle of the seat pan and the backrest.

The lever for adjustment 4 controls both functions at one time, and the seat pan and backrest cannot be set at independent angles. The consequence of the fixed angle between the seat pan and the backrest is that when the backrest is in its most vertical position, the seat pan is angled downward at the front and users feel like they have slid forward. If the adjustment is set so that the seat pan is more horizontal, the backrest is positioned in a rear-reclining position and reduces back support. The seat pan does lower below 17.5 in.

| Backrest vertical, seat pan front low. | Seat pan flat, backrest angled back. | Seat pan rear tilt, back very reclined. |

During the evaluation, nearby employees mentioned that they have also tried to switch chairs because even when they tilt the seat pan forward, they do not receive sufficient upper back support from the new chair. When any chair backrest does not provide support to the upper back, users must use their own muscles to remain in a vertical position, which increases fatigue and will often cause a user to lean forward to support the upper torso with his or her arms on the desk.

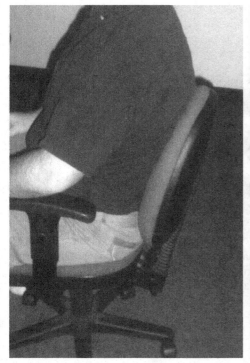

Co-worker in the old-style chair.

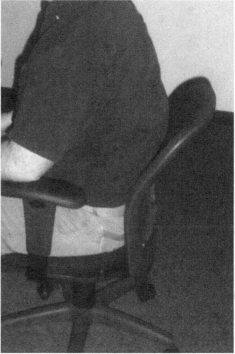

Co-worker in the new chair.

Objectives

The new chair is unsatisfactory because (1) it does not go as low as she needs it to go (16.5 in.), and (2) the forward seat pan angle will require support from her legs when the backrest is straight, and if she makes the seat pan flat, the backrest provides support to only a very small part of her lower back.

Recommendation 1

Allow her to work in a chair other than the new chair.

Outcome Expected Ms. Bayles reports of discomfort are clearly affected, and then resolved, by the type of chair she is sitting in. During the evaluation, her discomfort was mitigated by sitting in the old chair and the demo chair, which is available for special situations. She states that sitting in the old chair for many years had not been problematic to her preexisting condition, even with her long workdays.

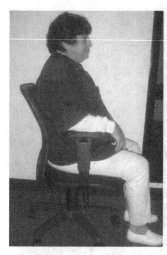
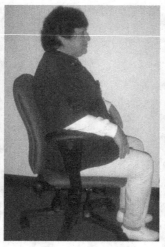
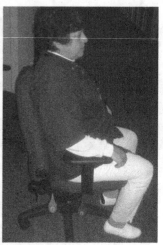
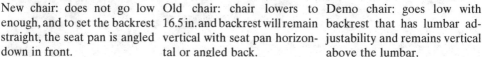

New chair: does not go low enough, and to set the backrest straight, the seat pan is angled down in front.

Old chair: chair lowers to 16.5 in. and backrest will remain vertical with seat pan horizontal or angled back.

Demo chair: goes low with backrest that has lumbar adjustability and remains vertical above the lumbar.

The chair that Ms. Bayles uses for her work should have the following three features:

1. The seat pan lowers to 16.5 in. from the floor.
2. The chair has independent controls for the backrest angle and the seat pan angle.
3. The chair has good lumbar support that is adjustable for its height and, ideally, its depth.

Note: The new chair is also deficient in that the backrest does not have any vertical adjustment, and it is likely that more than one employee observed during the evaluation will exceed the weight limits of a chair like this, especially for 24/7 use.

16. ARTHUR CURTIS: *SHOULDER*

History

Arthur Curtis is right-handed, 5 ft 8 in. tall, and has worked as an associate scientist at Acme for two years. He has been in his current work area for approximately 17 months and has been experiencing a recent onset of discomfort in his right shoulder. He reports that he has sought medical treatment during the last seven years for right shoulder discomfort, which he states was diagnosed as a problem in his right rotator cuff. He also reports that he has previously been diagnosed with right carpal tunnel syndrome, which was resolved through conservative treatment. He states that symptoms have recently been aggravated by a special project that requires much more time at his computer than in the lab.

Observations

Mr. Curtis works at a modular corner workstation with the work surfaces set to 29 in. from the floor. His keyboard and mouse are placed on the work surface and his monitor is placed on his CPU. He sits in a fully adjustable task chair that he has adjusted to 20 in. from the floor. The depth-adjustable seat pan is adjusted to its fully retracted position, and he demonstrates a sitting posture where his legs are moved to the outside of the seat pan, and he places his feet on the chair legs. He works with a waist-mounted pack that he places in front of him, which requires him to add a distance of 6 in. between the desk edge and his torso.

As he views his monitor, he is observed to tilt his head back, with his neck in a moderately extended posture, which is not recommended. He reports that he will soon be getting progressive lens glasses that will place his computer viewing lens toward the bottom of the lens. The progressive lens glasses will probably increase his neck extension as he views his monitor through the lower lens. His preferred viewing distance to his monitor is 24 in., and he is observed to lean forward as he views it because his monitor screen is 22 in. back from the desk edge and he is moved back 6 in. by the thickness of his waist pack.

Recommendation 1

Place the keyboard and mouse on an articulating keyboard tray.

Outcome Expected   To allow him to use his keyboard and mouse with his wrists at, or below, his elbow height, his keyboard and mouse were placed on an articulating tray. It was adjusted to 26 in. from the floor, which is 3 in. lower than it had been and places his wrists at a comfortable position below his elbow height. The tray was also adjusted into a slight negative tilt to maintain a neutral wrist posture. This allows him to move his elbows back, his shoulders into a more neutral posture, and to be able to sit back to receive support from the backrest of his chair. He is advised to remove the waist pack when working at his computer so that he can bring the keyboard and mouse closer to his torso.

Recommendation 2

Lower the monitor 3 in. and move it close to the desk edge.

Outcome Expected   To eliminate his neck extension when viewing the monitor, the monitor was moved off his CPU and placed on the desk. That height (4 in. lower) was a bit too low for a neutral neck posture, so it was placed temporarily on a 1-in.-thick book. When he receives his new progressive lens glasses, where his viewing lens will be at the bottom of the lens, he will probably want to remove the book so that the monitor is lower still. If he needs to lower it further, the swivel base can be removed easily.

Recommendation 3

Speed up the "motion" of his mouse.

Outcome Expected   To reduce his right arm and wrist motions to operate his mouse, he was shown how to increase the "speed" of his mouse in the Windows control panel. The mouse speed was increased from 50% to 75%, and he will modify this as he becomes accustomed to its functioning and to accommodate other mousing devices that will be reviewed.

Recommendation 4

Learn keystroke alternatives to mouse clicks to reduce mouse use.

Outcome Expected   To reduce his movement between the keyboard and the mouse, he was shown a comprehensive list of keystrokes that replace mouse clicks. He printed the 35-page list that is available at the evaluator's website.

Recommendation 5

Try using an angled keyboard with an embedded touchpad.

Monitor on CPU, with keyboard and mouse so high that his elbows are forward and his head is tilted back.

Monitor lowered to 1-in. book for neutral neck. He sits back with a touchpad keyboard on a tray.

Outcome Expected   To prevent him from having to move his right arm between his mouse on the right and the keyboard keys, he was given a keyboard that has an embedded touchpad as a mousing device. This was immediately beneficial, and he is accustomed to using a touch pad in a similar posture on his laptop.

17. PAT CARTER: *SHOULDER*

History

Pat Carter is right-hand dominant, 5 ft 6 in. tall, and has worked as a senior project manager at Acme for 10 years. She has been in a new cube for five weeks and reports shoulder discomfort as she works at her computer.

Observations

Ms. Carter works at a modular corner workstation with the work surfaces set 29 in. from the floor. Her keyboard and mouse are placed on the work surface and she uses a large angled keyboard. She has her laptop computer placed on a 6-in.-tall riser behind her keyboard, and she uses the laptop screen as her monitor.

She states that she uses the laptop screen because large CRTs cause eye fatigue and she is concerned about the radiation from a CRT. Additionally, she states that she prefers to have her screen lower than her eye level.

She sits in a semiadjustable task chair that has adjustments for height and tilt tension only. The seat pan is raised to 21 in. from the floor, which is relatively high for her, and the tilt tension is set moderately low. This height is used to elevate her arms up to desk height. As she sits in the chair with her arms at her side and her hands in her lap, her seated elbow height is 27.5 in. from the floor. As she types, she is observed to lean forward in her chair and rest her forearms on the desk surface, which rounds her shoulders and thoracic spine and contributes to her discomfort. This is a typical posture that results when the keyboard and mouse are higher than a user's seated elbow height. Because her keyboard has a rise in the center and she has the rear legs extended, her hands are almost 3 in. higher than her seated elbow height.

There is an articulating keyboard tray installed under her computer surface, but she does not use it because she likes to have her keyboard positioned in a slight positive tilt, and the keyboard tray is made for a standard keyboard.

Initial: elbows forward with rounded shoulders.

Final: upper arm against torso, shoulders back.

Recommendation 1

Tighten the tilt tension of the chair.

Outcome Expected   At the time of the evaluation, the tilt tension on her chair was tightened from approximately 30% to 75%. This allows her to lean back and receive support to her back so that her muscles do not have to be engaged for her to sit upright as she works. The lack of sufficient tilt tension will often cause a person to lean forward, as she does, because the chair will recline when she leans back into the backrest for support.

Recommendation 2

Install an external monitor that is smaller than the 21-in. versions.

Outcome Expected   Because the small size of her laptop screen may contribute to her tendency to lean forward as she views her computer data, she is advised to work with an external monitor. Because she reports eye fatigue from a CRT and has been accustomed to working with a 12-in. laptop screen, she should work with a monitor that is smaller than the 21-in. models used elsewhere in her area. Also, changing the refresh rate of a CRT monitor to an odd interval such as 75 Hz will reduce the interference noticed by some, as a result of proximity to fluorescent lights, which produce a normal 60-Hz flicker.

The evaluator has conducted detailed studies with the University of Colorado on large CRTs at computer stations. From extensive research and observations of typical situations presented by the evaluator, their findings determined that the health effects of proximity to CRT monitors were extremely low.

Recommendation 3

Move the keyboard and mouse onto the keyboard tray.

Outcome Expected   To reduce her forward lean and keep her shoulders back, using her more supportive backrest, her keyboard and mouse were moved to her keyboard tray during the evaluation. Because her keyboard tray appeared to be designed for a standard keyboard, as reflected by the wrist rest only, the evaluator removed the wrist rest to reveal that the platform was actually designed for an angled keyboard such as hers. With her wrists at her seated elbow height, she will be able to sit back more easily, with neutral postures.

Wrist rest is straight and fits a standard keyboard.

Removing the wrist rest reveals an angled platform.

18. ROSE KRAMER: *SHOULDER*

History

Rose Kramer is right-handed, 5 ft 3 in. tall, and has worked as a quality systems administrator at Acme for five months. She has been at her current workstation for one month and reports discomfort in her right shoulder that began upon moving to this area.

Observations

Ms. Kramer sits at a corner section of the 28-in.-high work surface and places her keyboard and mouse on the surface. She sits in a fully adjustable task chair and she has the seat pan set 18 in. from the floor, which places her hips slightly above her knee height at an optimal lower extremity posture. Her resulting seated elbow height is 25 in. from the floor, which is 3 in. lower than her work surface heights.

As she types, her elbows are well below her wrists and cause her to rest her wrists against the front edge of the work surface. This is an awkward posture because of the contact pressure of her wrists against the hard surface and from having to elevate her hands above elbow height. Studies have shown that when performing computer work with the wrists above seated elbow height, the shoulders are drawn forward, which Ms. Kramer demonstrates.

To operate her mouse she extends her right arm forward to raise her right elbow high enough to accommodate the work surface height, and as a result, she rounds her right shoulder as she extends forward. In this posture she is pulled forward and does not receive any support from her chair's backrest. To provide some upper extremity support in this posture, she will lean on her left forearm so that both forearms are resting on the work surface, incurring significant contact pressure against the tissues of the underside of her forearms and elbows. Because she does not use wrist rests, she demonstrates approximately 20° of right wrist extension (bending back) as she uses her mouse. When typing, her wrists are in a flexed posture.

Her chair has the lumbar support at its lowest position and the backrest is unlocked, with the tilt tension at its lowest setting. The depth-adjustable seat pan is at its deepest (extended-out) position, which prevents her from moving all the way back in her chair to use the backrest. Additionally, with the tilt tension low and the backrest unlocked, the chair will recline if she rests against it. Therefore, to sit in the chair as it is, she must engage her muscles to support her torso, since the chair is not pushing against her.

Mouse arm extended with right shoulder rounded.

Keyboard/mouse on surface with wrists above elbows.

Recommendation 1

Adjust the chair for optimal back support and seat pan depth.

Outcome Expected At the time of the evaluation Ms. Kramer was shown how to adjust the lumbar support height, the seat pan depth, and how to lock the backrest into an upright position. These features were adjusted for her, and they made a significant improvement in her comfort. Having the chair adjusted for her will allow her shoulders to relax, especially as Recommendation 2 is invoked.

Recommendation 2

Install an articulating keyboard tray for the keyboard and mouse.

Outcome Expected   To allow her to work with her wrists at or slightly below her seated elbow height with her upper arms parallel to her torso, her keyboard and mouse should be placed on an articulating tray. Having her wrists closer to her elbow height will allow her to sit back in her chair, with a neutral arm and shoulder posture. Additionally, the keyboard tray will have a wrist rest that will position her wrists in a neutral posture with a cushioned surface under them.

19. FLORA GIBBS: *SHOULDER, LOWER BACK*

History

Flora Gibbs is right-handed, 5 ft 8 in. tall, and has worked in her current workstation for eight months. She has been experiencing discomfort in her lower back and in her shoulders at her neck R > L.

Observations

Ms. Gibbs works in a corner modular workstation with the work surfaces set 29 in. from the floor. Her angled keyboard and mouse are placed on the work surface. She sits in an adjustable chair with a tall backrest that is set at its lowest position and therefore does not provide upper back support. This chair positioning places her seated elbow height 26.5 in. from the floor, which is 2.5 in. lower than the work surfaces. Working with her wrists that much higher than her seated elbow height causes her to move her keyboard and mouse back on her desk and to rest her forearms on the desk surface with her elbows abducted out from her torso, which is not recommended.

Recommendation 1

Raise the backrest so that it aligns with her spine.

Outcome Expected   Her chair's backrest was raised almost 5 in. to provide support to her upper back while still providing support to her lower back. This will invite her to sit back with her scapulae against the backrest and thereby reduce the muscle engagement needed to sit upright, which will reduce the strain she feels in her shoulders.

Recommendation 2

Install an articulating keyboard tray for an angled keyboard.

Outcome Expected　To allow her to work with her wrists at, or slightly below, her seated elbow height and her arms parallel to her torso, her keyboard and mouse should be placed on an articulating keyboard tray. She tried the articulating keyboard tray of a co-worker during the evaluation, and found it very comfortable and that it positioned her upper extremities in neutral postures.

Initial: backrest low and elbows out.　Backrest raised for upper back.　Keyboard tray creates neutral arms.

20. JOE CASWELL: *SHOULDER, LOWER BACK*

History

Joe Caswell is right-handed, 5 ft 10 in. tall, and has worked as a project manager at Acme for seven months. He has been at his current cube for three months and has been experiencing discomfort in his lower back, just above the buttock on the right side. He reports that the discomfort began as intermittent episodes in late January and by the end of March had become a daily occurrence, with pain running down his right leg into his calf. He also states that on Sunday and Monday preceding the evaluation, the pain was extreme and his productivity at work was affected significantly. He is receiving medical attention for this. He also reports discomfort in his right shoulder that began about two weeks before the evaluation.

Job Activities

Computer and phone work

Sitting stationary

Static wrist and arm postures

Observations

Mr. Caswell works at a modular corner workstation with the work surfaces set 29 in. from the floor. His standard keyboard, optical mouse, and flat-panel monitor are placed on the work surface.

He sits in a small task chair that he just started using because the cushioning felt better than his previous chair and the seat pan was not as wide. He elevates the seat pan to 20 in. from the floor, which correctly places his hips slightly above knee height; however, he has a long upper leg, and the chair seat pan does not extend far enough along his leg to support him properly. He is observed to sit with a leg posture where his right foot is turned outward and he rests his heel against the outside of the chair leg.

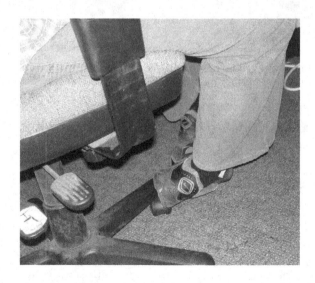

Although his discomfort reports are in his lower extremities, he does demonstrate rounded shoulders and a somewhat hunched posture, especially with the monitor so low. With his chair at an optimal height of 20 in. from the floor, his seated elbow height is 26 in. from the floor, which is 3 in. lower than his work surfaces. Because keyboard and mouse activities should be performed at, or slightly below, seated elbow height, Mr. Caswell uses several tactics to elevate his elbows. One is that he sits back from his keyboard, which moves his elbows forward and up, and a second is that he moves the chair armrests outward, which also elevates his elbows.

Although these postures position his elbows at his working wrist height, they also create shoulder flexion and rounding, which is not recommended. He states that he prefers using an angled keyboard, as he does at home. His posture of moving his elbows out supports an angled keyboard, but that posture should change.

His backrest does not provide upper back support and with the monitor low and his keyboard/mouse above seated elbow height, his shoulders are rounded.

He will often work from reference documents that he places to the left of his keyboard, causing an awkward turn and bend to read them. Right shoulder is forward.

Recommendation 1

Replace the chair with a chair that has a deeper (or adjustable) seat pan and a tall backrest.

Outcome Expected To provide more support to his upper legs and upper back, Mr. Caswell was allowed to try the nonstandard demo chairs in the human resources department.

His current chair does not extend far enough along his upper leg to provide sufficient support.

When sitting in a chair that has a sliding seat pan, the seat pan can be positioned for optimal leg support.

Recommendation 2

Install an articulating keyboard tray for the keyboard and mouse.

Outcome Expected To allow him to place his wrists at, or slightly below, his seated elbow height, he tried a keyboard tray in the human resources demo area. The tray worked very well for him because it places his shoulders in a very neutral posture and allows him to sit back against the chair backrest. Accordingly, he will benefit from an articulating keyboard tray being installed at his desk.

His seated elbow height is 26 in. and the desk height is 29 in., causing him to extend his arms and round his shoulders.

With a keyboard tray he can position his keyboard and mouse at his seated elbow height and then maintain a neutral upper extremity posture.

Recommendation 3

Try using a footrest.

Outcome Expected Because he is inclined to use differing foot postures, and to help him avoid the posture of turning his right foot out,
he was asked to try using a footrest from the demo area. If he finds that
this helps his discomfort (even in his current chair), he should acquire
one.

Recommendation 4

Acquire a wide document holder to position between the keyboard and
the monitor. The monitor should also be raised.

Outcome Expected Because he often uses an awkward posture when
viewing reference documents to the left of his keyboard, and will have
his keyboard off the desk surface (on a tray), he would benefit from a
document holder placed between his monitor and keyboard. He tried
the demo version in the human resources department, which is the ideal
device for him to use. It is only 14 in. wide; he should use one that is
18 in. wide.

Recommendation 5

Replace his standard keyboard with an angled keyboard.

Outcome Expected   Because he is accustomed to using an angled keyboard and he demonstrates moderate ulnar deviation in his wrists as he types on a standard keyboard, he would benefit from an angled keyboard to keep his wrists in a neutral (straight) posture. The photo demonstrates 30° of ulnar deviation in right wrist.

21. ETHEL HOLLIS: *SHOULDER, NECK*

History

Ethel Hollis is right-handed, 5 ft 6 in. tall, and works as a building manager. She has been experiencing discomfort in her neck and shoulders as she works in her office. She works at a 29-in.-tall desk that has a left-side "secretary return" that was designed for a typewriter.

Observations

She positions her computer station at the left corner of the main desk surface and positions her CPU and flat-panel monitor in the corner, close to the walls. Her keyboard is positioned in front of her monitor and to the right, so that it rests on the taller desk surface.

She sits in a semiadjustable chair with a medium-sized backrest, and because the armrests hit the edge of the desk on each side, she cannot move her torso in close to the keyboard. This causes her to reach forward for her keyboard and mouse and prevents her from receiving support from her chair backrest.

Ms. Hollis has her phone placed on the right side of her PC and she is observed to adduct her left arm across her torso to pick up the receiver. She will often conduct phone conversations while she is accessing information on her computer and cradles her phone receiver between her neck and left shoulder, which probably contributes to her shoulder discomfort on both sides.

Sitting forward to reach keyboard and mouse; no back support.

Extended reach and turn to acquire the receiver.

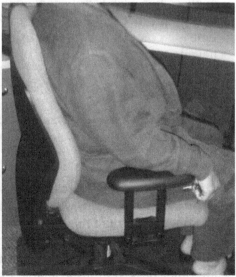

Medium backrest where the top angles away from shoulders.

Tall-back chair that remains vertical at the top.

Recommendation 1

Replace the semiadjustable medium back chair with a fully adjustable tall-back chair.

Outcome Expected   To allow her to adjust the angle of her chair's seat pan and to have upper back support (as well as improved cushioning), she tried sitting in a fully adjustable tall-back chair like those that some employees use. It provided significantly improved upper and lower extremity support, and a building representative brought her one that was not currently in use.

Recommendation 2

Move the flat-panel monitor to the center of the desk.

Outcome Expected   To allow her to work in a straightforward orientation to her desk, her monitor should be moved from the CPU on the left to the center of her desk, directly in front of her. This will allow her to move her keyboard and mouse where they will not require her to reach for them or to sit on the front edge of her chair. This will lower the monitor 6 in. and she should use a simple monitor riser to raise it approximately 4 in. above the desk height. This is important so that the angle of her neck is not changed dramatically, since a dramatic change in neck posture will frequently cause discomfort. A change of 2 in. at a time is recommended for approximately one week at the newly changed height.

Recommendation 3

If possible, replace the desk and secretary return with a work surface or desk system that can be positioned between 27 and 28 in. from the floor.

Outcome Expected   Because her seated elbow height was 27 in. in her original chair and she will probably place the seat pan of a tall-back chair no higher than that, having her keyboard and mouse at 27 to 28 in. will be ideal. This is so that her wrists will be at, or below, her seated elbow height as she works on her keyboard and mouse. Because there are other desk systems available that can be adjusted or are lower, she will try to acquire one. If she finds a suitable desk that is taller than 28 in., she should have an articulating keyboard tray installed so that she can place the keyboard and mouse 27 to 28 in. from the floor.

Recommendation 4

Move the phone closer and use a hands-free headset.

Outcome Expected To reduce her reach for her phone and prevent her from cradling her phone receiver in her shoulder as she talks while working on her computer, she should acquire a hands-free headset. Additionally, moving the phone closer to her will reduce the extended reach for dialing (done with her right hand), even though she will not use the receiver when she uses the headset.

22. MARIE HOLT: *SHOULDER, NECK*

History

Marie Holt is right-handed, 5 ft 4 in. tall, and has worked as a client services representative at Acme for two years. She has been in her current work area for one year and has been experiencing discomfort in her right shoulder and neck for the last month. She reports that her discomfort began without a recognizable incident and that during a recent three-day leave, her discomfort diminished significantly.

She had a previous ergonomic evaluation in response to neck and shoulder discomfort experienced after working for extended periods. She reported at that time that she injured her thoracic spine in a motor vehicle accident eight years earlier and had undergone conservative treatments for discomfort.

Job Activities

Sitting
Computer input activities: mousing and keyboarding
Repeated phone conversations
Handwriting

Observations

Ms. Holt works at a modular corner workstation that has independent crank-adjustable work surfaces for her input devices and her monitor. The front section that holds her keyboard and trackball is positioned 27 in. from the floor. As she types, she has her wrists slightly lower than her elbows, which is recommended. Her flat-panel monitor is height-adjustable and is at its lowest position, sitting on the independent monitor surface, that is adjusted to 30.5 in. from the floor. She is observed to bend her neck to look down at the lower sections of the screen because she wears bifocal glasses and views her monitor through the upper lens.

She sits in an adjustable chair that she has set 18 in. from the floor, and the height-adjustable backrest is at its lowest position. The backrest is tilted slightly rearward, with the armrests set 28 in. from the floor. She is observed to sit forward of her chair backrest, which is partially caused by the rearward angle of the chair backrest. Additionally, the backrest is at its lowest position, so that the lumbar protrusion is below her lumbar lordosis (inward curve). That positioning aligns the lumbar support against her hips, moving her forward and creating a gap between the backrest and her lumbar curve. She has tried to place a small cushion in this area, but it does not stay in place.

Her phone is positioned to the right of her keyboard, on a 1-in.-thick book 10 in. back from the front edge of her desk. To operate the keys on her phone, she must fully extend her arm outward and abduct it to the right. This motion is the movement that aggravates her discomfort more than any other work task, and she is observed to grimace and rub her shoulder each time she performs this reach.

Phone far right, arm extended, shoulder abducted.

Phone in center, no abduction, use by both hands.

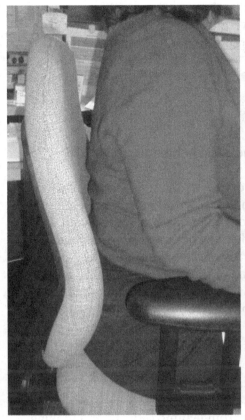 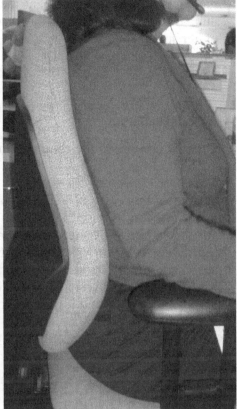

Initial: backrest low and angled back.   Final: backrest raised and angled forward.

Objectives

The primary objectives for Ms. Holt are (1) to reduce the reach for her phone, which is the action that causes her the most discomfort, and (2) to increase the chair support to her back and shoulders.

Recommendation 1

Adjust the chair for shoulder support and an upright posture.

Outcome Expected   To place her lower extremities in comfortable neutral postures, the seat pan of her chair was raised from 18 in. to 19 in. from the floor. The cushion she had on her backrest was removed and the backrest was raised so that the lumbar protrusion aligns with her lumbar lordosis and allows the curve of the chair to follow the curves of her spine. The backrest was also repositioned to a more vertical orientation so she can sit upright and still have full back support as she works.

Recommendation 2

Move the phone to the center of the desk, behind the keyboard.

Outcome Expected   At the time of the evaluation, her phone was moved to the center of her desk and placed on ½-in. spacers, and the work surface was raised 1 in. to match the increase in chair height. This positioning prevents her from abducting her right arm beyond the centerline of her torso and keeps it lower. This positioning also allows her to use her left hand to press some of the phone keys. Although she was initially uncertain of using her left, nondominant hand; she readily became adept at using it for all phone functions.

Recommendation 3

Raise the monitor to create a neutral neck posture.

Outcome Expected   To prevent her from leaning forward and looking down to see the lower sections of her screen, her 19-in. monitor was raised so that her horizontal line of sight is one-third of the way down from the top of the screen. The optimal height for her is when the bottom of the screen is 9.5 in. above the 28-in.-high work surface: a 37.5 in. total height from the floor to the bottom of the screen.

Recommendation 4

Move the monitor closer and move it into a level position.

Outcome Expected   To reinforce her upright posture and prevent a forward lean as she works, her monitor was moved approximately 3 in. closer to her. Her monitor was measured to be ½ in. higher on the right than on the left, and although this is easily modified, she has become accustomed to this positioning. Because this could have a slight effect on her discomfort, she should make it level, and it was corrected during the evaluation.

Final Settings

Chair seat pan height: 19 in.

Chair armrest height: 28.5 in.

Work surface height for keyboard and mouse: 28 in.

Monitor height (to bottom of screen): 37.5 in.

23. PAULA DIEDRICH: *SHOULDER, NECK*

History

Paula Diedrich is right-handed, 5 ft 4 in. tall, and has been experiencing significant discomfort from the base of her neck (posterior) into her shoulder and down into her right upper arm. She reports that her symptoms have been increasing for the last few weeks and that she believes they were triggered by an increased workload that began with a new assignment six weeks prior to the evaluation. She states that her discomfort subsides significantly during her time away from work.

Observations

She works at a computer workstation that has an adjustable surface for her keyboard and mouse, and she does not use wrist rests. The adjustable height surface is set at 27.5 in. and her flat-panel monitor is placed on the fixed height surface 29.5 in. from the floor.

She sits in a semiadjustable task chair with the seat pan set 18 in. from the floor and the backrest at its lowest position and angled back. With the backrest so low, the lumbar protrusion aligns with her hips, which pushes her forward on the chair and creates a gap at her lumbar lordosis (inward curve). This backrest positioning prevents her from receiving any back support from the backrest, especially as she leans forward with her arms extended as she works. This chair height places her seated elbow height at 26 in. from the floor, which is 1.5 in. lower than the keyboard and mouse surface, and her horizontal line of sight meets the monitor at the very top of the screen.

Although this seat pan height of 18 in. is optimal for positioning her lower extremities in neutral postures, with her hips slightly higher than her knees, it causes her to elevate her wrists higher than her seated elbow height as she works at her computer, which is not recommended. She demonstrated that the adjustable keyboard and mouse platform seems to be broken, and she has been unable to move it.

She regularly works from reference documents which she places in front of her keyboard and/or on her lap. When viewing those documents she is observed to flex her neck and tilt her head down to view them and then to extend her arms outward to reach the keyboard and mouse. She also abducts her right arm outward as she operates the mouse, even when she is not working with documents.

Document position causes significant neck flexion.

Leaning forward and extending arms for keyboard and mouse.

Because the most egregious postures observed were flexing her cervical spine as she looks down at documents and extending her arms for her keyboard and mouse, she was asked to try a document holder that positions documents between the keyboard and the monitor. She was also asked to try using a mousing device that positions her hands directly in front of her keyboard home keys and allows her to use either hand for cursor movements and clicking. This positioning brings her right upper arm back so that it is parallel to her torso and not abducted.

Recommendation 1

Lower the keyboard mouse platform to place her wrists at, or slightly below, seated elbow height.

Keyboard platform lever is broken but could be lowered 2.5 in. Still, her right arm is abducted 35°.

Backrest raised, a document holder added, and a mouse device directly below her keyboard home keys.

Outcome Expected   The lever that allows the user to raise and lower the keyboard platform is broken, but the evaluator was able to lower it 2.5 in., to a height of 25 in. from the floor, so that her working wrist height is slightly below her seated elbow height.

Recommendation 2

Raise the chair's backrest and move it forward, closer to the chair seat pan.

Outcome Expected   To allow her to receive some middle and upper back support from her chair, the backrest was raised from its lowest position to its highest position, and it was moved forward approximately 1.5 in. These modifications invited her to rest against it more readily, especially when using the mousing device, which allows her to keep her arms at her side.

Recommendation 3

Try working with a mouse device that is positioned directly in front of the keyboard space bar and that can be operated with either hand.

Outcome Expected To allow her to operate her mouse with her hands and arms closer to her torso, she was allowed to try a mouse that is positioned directly in front of the keyboard space bar. She was very adept at acclimating to this new input device, which allows her to use either hand and keeps her arms close to her torso. She felt that this led to a significant improvement in her shoulder and arm comfort.

Recommendation 4

Try working with a document holder between the keyboard and the monitor.

Outcome Expected To reduce her cervical flexion from looking down at reference documents, an 18-in.-wide document holder was positioned between her keyboard and monitor. This was very beneficial because it is supported by a 1-in.-tall monitor riser that elevated her monitor 1 in., placing her eyes approximately one-third of the way down from the top of the monitor, which is optimal for her.

24. RACHEL SIMPSON: *SHOULDER, NECK*

History

Rachel Simpson is right-handed, 5 ft 6 in. tall, and an attorney who has worked at her current workstation for two years. She reports that she has been experiencing neck pain and shoulder pain R > L for the last four months. She states that she has slight scoliosis in her cervical spine, which contributes to her neck discomfort.

Observations

Ms. Simpson works at a standard 28.5-in.-high double-pedestal desk with her computer positioned on a separate nonadjustable computer table. She works with many hardcopy documents that she will place on her desk, to the left, for computer work, as well as on her desk surface for reading. Both activities create significant neck flexion and also a left turn in her neck, with repeated neck movements when she is working from documents while at her computer.

She has been trying a variety of chairs that are used in other offices but has not found a comfortable chair from those available. During the evaluation, she was using a fully adjustable mesh chair that had not been adjusted for her other than for chair height.

Turning neck, looking downward for documents.

Neck flexed, shoulders rounded for reading.

Neck turn is significantly reduced by document holder between keyboard and monitor.

Neck flexion reduced with elbows and shoulders back into a more neutral posture with document holder.

Recommendation 1

Adjust the lumbar support and tighten the tilt tension in the chair.

Outcome Expected The lumbar support of the chair that she felt was most comfortable was aligned with her lumbar lordosis (inward curve) and the tilt tension was tightened from approximately 25% to 75%. The spinal alignment of the lumbar support and the enhanced back support from the increased tension will allow her muscles to relax as she sits upright for computer work.

Recommendation 2

Use a document holder above the keyboard and on the desk.

Outcome Expected With her chair configured to provide more back support, she was allowed to try document holders that are 18 in. wide and are sturdy enough to allow her to write on them. These document holders, at her computer and for reading, reduce the angles and motions significantly.

25. STEVE BRAND: *SHOULDER, NECK*

History

Steve Brand is right-handed, 6 ft 1 in. tall, and works as a team leader in the Acme Information Services Group. He has worked at his current desk for three years, works 40+ hours per week, and estimates that he spends 60% of his workday at his computer. He states that he does not experience significant discomfort other than stiffness in his neck as it meets his upper shoulders on each side. However, he reports that it has been getting increasingly worse lately.

Observations

Mr. Brand works at an old wooden double-pedestal desk that is 30 in. tall. Because of his height, he has his semiadjustable task chair seat pan set 23 in. from the floor, and his arm rests 30 in. from the floor. This chair positioning places his lower extremities and arms in neutral postures.

His keyboard, mouse, and monitor are placed directly on the desk, about 32 in. from his eyes. He uses an old wrist rest for his keyboard but does not have a wrist rest for his mouse and places the mouse 16 to 20 in. back from the front edge of the desk. This mouse positioning causes him to fully extend his right arm as well as bending it back in moderate wrist extension.

Mr. Brand places his mouse in a far-forward position because his phone is where the mouse is typically placed. He will often adduct his left arm across his torso to pick up the receiver. Similarly, he has his day-timer to the left of his keyboard and must adduct his right arm across his torso to write in it. Additionally, his reach for the mouse contributes to pulling him forward, away from his chair backrest.

He wears progressive lens glasses and uses the lower sections of the lens for viewing his screen. Although he uses the lower lens, his monitor is still too low and he must tilt his head down to view his computer screen. This head and neck posture tends to pull him forward away from his chair backrest.

Monitor low; 25° of neck flexion and "forward head."

Monitor on 3.5-in.-thick phone book; 15° of neck flexion.

Recommendation 1

Elevate the monitor to reduce his neck flexion.

Outcome Expected   To reduce his neck flexion, his monitor was placed on a 3.5-in.-thick phone book, which reduced his neck flexion by 10°. This adjustment also induced him to sit back in chair and receive increased upper back support. It appears that his monitor should be elevated further; however, monitors should not be raised more than approximately 2 to 3 in. at a time, because a dramatic change in monitor height at one time will often create neck and/or shoulder discomfort.

He will purchase an adjustable-height monitor riser. He states that his current eyeglass prescription is two years old, and before he makes a permanent adjustment to his monitor height, it would be beneficial to have his prescription updated in case that causes him to view his screen through a different part of the lens.

Recommendation 2

Switch the positions of the phone and the day-timer.

Outcome Expected   To eliminate his reach across his torso for his phone and his day-timer, their positions were reversed. Because this will prevent him from leaning forward as much to use these items, he will be more inclined to sit back in his chair and let his muscles relax from the increased back support he receives.

26. LAURA TEMPLETON: *SHOULDER, UPPER BACK, KNEE*

History

Laura Templeton is left-handed, 5 ft 7 in. tall, and has worked at Acme for five years. She has worked in her current office for that entire time and has been experiencing discomfort in her knees R > L, in her middle back, and in her right shoulder. She reports that her knee discomfort sets in after prolonged sitting and that she has been receiving weekly physical therapy treatments for the last month. She also has discomfort in her midback and her right shoulder (above her scapula) that developed shortly after she began working in her current office. She did not, initially, address the shoulder discomfort, thinking it was expected from the work she does.

Observations

Ms. Templeton works at a height-adjustable corner-oriented work surface that is positioned 28.5 in. from the floor. Her standard keyboard and mouse are placed on the work surface and she does not use wrist rests. She positions her keyboard and mouse approximately 12 in. back from the edge of the desk and rests her forearms on the desk surface as she types and operates her mouse. Her height-adjustable flat-panel monitor is placed on the desk surface at is lowest position, and her horizontal line of sight meets the screen at the very top of the monitor. She reports that she will occasionally slouch, with her head lowered and her shoulders flexed.

She sits in a standard semiadjustable task chair that has a small backrest that is curved from side to side, and the seat pan has only minimal angle adjustability. She positions the seat pan height slightly below her knee height so that she can extend her right leg forward to widen the angle at her knee as her right foot rests on the floor. She will often rest her right foot on her heel when it is extended forward, which causes some heel discomfort.

This seated height places her elbows 26.5 in. from the floor, which is a few inches below the height of her computer work surface. With her wrists higher than her seated elbow height, she moves her keyboard and mouse back from the front edge of the work surface so that her elbows rise to her wrist height. This posture also moves her forward of her chair backrest, reducing the support to her middle and upper back, above her lumbar spine.

Arms are extended with shoulders raised to elevate elbows to desk height.

No upper back support, and low monitor position draws her forward.

Objectives

The primary objectives for Ms. Templeton are (1) to replace her small chair with one that provides more leg support and upper back support, (2) to modify the angle of her hips, upper legs, and knees as she sits by lowering the front of the chair seat pan, (3) to lower her wrists to seated elbow height, and (4) to raise her monitor, especially after lowering the work surfaces.

Recommendation 1

Replace the chair with a fully adjustable chair that has full back support and a seat pan that can be angled down at the front.

Outcome Expected   At the time of the evaluation, Ms. Templeton tried sitting in some of the fully adjustable chairs that the EH&S department has for special situations. She found that a tall-back, mid-sized, fully adjustable chair was the best fit. The backrest, with adjustable lumbar depth, aligns very well with her spine, and the seat pan can be angled into a slight forward tilt to open the angle at her hips and knees. She will try this chair for a few days to measure its benefit to her comfort.

Recommendation 2

Try a wedge-shaped cushion as a footrest, and raise the chair slightly.

Outcome Expected   Since she needs to extend her right foot at times, she was given a chair cushion that is wedge-shaped to use as a footrest. This cushions her foot (especially, when resting on her heel) and allows her to sit a bit higher so that her hips are above her knee height. She is also encouraged to work with her shoes off for circulation and ankle movement.

Recommendation 3

Position the work surface at, or slightly below, seated elbow height.

Outcome Expected   To allow her to work in neutral upper extremity postures, the work surfaces were lowered to just above her seated elbow height. Ideally, they would be lower than her seated elbow height, but the change performed during the evaluation was dramatic and she wishes to make this change incrementally over the next few weeks. Even with the work surfaces moderately lowered, she is able to pull her keyboard and mouse closer to her for a more neutral upper arm and shoulder posture as she engages the tall backrest at her middle and upper back.

Recommendation 4

Consider using a smaller keyboard to reduce her right arm reach to the mouse.

Outcome Expected Her current keyboard is 19.5 in. wide and the distance between the home keys and the mouse creates a significant shoulder motion as she moves her right arm back and forth. She tried a small keyboard that does not have a numerical keypad on the right, so that the mouse is closer.

The 19.5-in.-wide keyboard creates an extended reach to mouse and the home keys are positioned left of center.

Keyboard without a numeric keypad allows the mouse and home keys to be directly forward of torso.

Recommendation 5

Raise the monitor and move it a little closer.

Outcome Expected Since Ms. Templeton's initial monitor position had caused her to tilt her head down to view the lower sections, and her desk was subsequently lowered; her monitor needs to be raised so that her horizontal line of sight meets the screen one-third of the way down from the top. The monitor has slight height adjustability, but she also needs a monitor riser to attain the optimal height. It was also brought 2 in. closer to help ensure that she sits upright, engaging her chair backrest.

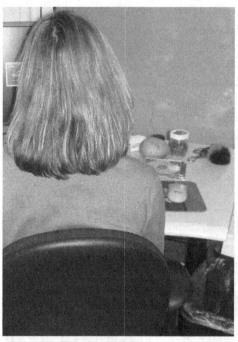 

Her current chair provides no upper back support and with her desk height above her elbow height, she leans forward.

A tall backrest and her mouse closer to center creates a neutral postures with back support; seat pan angled down.

27. FRANKLIN GIBBS: *UPPER BACK*

History

Franklin Gibbs is right-handed, 5 ft 11 in. tall, and has worked as a chemist at Acme for four years. He has been in his current work area for approximately 2.5 years and has been experiencing discomfort in his upper back just above his shoulder blades R > L. He reports that he was in a motor vehicle accident two years ago and has experienced intermittent back discomfort ever since. He states that during the last month his symptoms have increased, which he attributes to spending more time than usual at his computer workstation.

Job Activities

Sitting
Computer input activities: mousing and keyboarding
Handwriting and reading
Working with documents
Working in the lab (not observed)

Observations

Mr. Gibbs works at a 30 in. × 60 in. desk that is 29 in. tall. He positions his monitor on his CPU at the far left corner of the desk at an angle so that it does not usurp work space directly in front of him. His keyboard and mouse are positioned at an angle to match the monitor positioning and he extends his arms forward as he operates them, which also rounds his shoulders. He does not use wrist rests and demonstrates significant wrist extension (bending back) as he operates his keyboard and mouse.

He sits in a conference room chair that has adjustments for height and tilt tension only. He has it adjusted to 20 in. from the floor, which places his hips slightly higher than his knees and establishes a neutral posture for his lower extremities. The tilt tension is at its lowest setting and does not provide resistance to leaning back or support for sitting upright. As a result of the lack of tilt tension in his chair backrest, he must engage his own muscles to sit upright, or must sit in a forward lean, which puts a significant kyphotic curve in his upper spine. The curve is due partly to the fact that his chair's backrest does not rise high enough to provide any support to his middle or upper back if he could lean against it in an upright position.

To work on his computer at an angle to the desk, he will put one leg on either side of the left-side desk drawers. This leg position prevents him from moving in closer to his keyboard and mouse and contributes to his rounded shoulders. He prefers to maintain a 26-in. viewing distance from his monitor, which is created with him sitting forward, as he does.

Initial posture: arms extended with rounded shoulders.

Sitting straight; elbow height is lower than wrist height.

If he were to position his keyboard directly in front of him and sit upright, the current desk height would place his wrists higher than his elbows, which is not recommended. Although a keyboard tray, installed in the position of his desk pencil drawer would be a common solution, the space between the drawers on either side is 22 in., which is not wide enough for a keyboard and mouse to be positioned in the space.

His horizontal line of sight (sitting upright) meets his monitor at the very top of the screen and he will flex his neck and tilt his head down to view the lower sections of the computer screen. This also contributes to a "forward head" slouch as he tires and reaches forward for the keyboard and mouse.

He will often work from reference documents as he performs computer activities, and he places those documents on the available desk surface. Viewing those documents on the desk requires him to bend his neck forward and down, which also contributes to his forward head posture.

He does very little phone work or handwriting.

Objectives

The primary activities or postures to modify are to lower his wrists, have him work with his computer directly in front of him (not to the side), and sit in a chair that provides upper back support.

Recommendation 1

Sit in a chair that will provide upper back support.

Outcome Expected   To help him sit upright and to have support to his upper back, he was asked to try sitting in one of the fully adjustable task chairs that are used in the marketing department. This chair was very beneficial; he must move the depth-adjustable seat pan to its fully retracted position for the backrest to align with his spine.

Recommendation 2

Replace the 30 in. × 60 in. desk with a 30 in. × 72 in. desk with ample open leg space.

Outcome Expected   To allow him to work with his computer directly in front of him and have his wrists lower than his elbows, his current desk should be replaced with a version that has at least 28 in. of space for a keyboard tray below it. In the vacant office directly north of Mr. Gibbs's office there is a 30 in. × 72 in. table with more than 50 in. of unobstructed leg space on the center and right sides. To maintain a 26-in. viewing distance to his CRT monitor (which is 18 in. deep), the desk should be placed 5 in. away from the north wall and not up against it.

 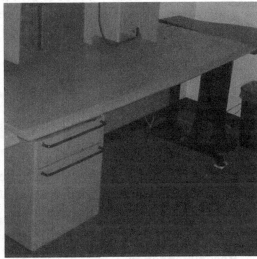

Sitting in one of the fully adjustable chairs.

Unused 30 in. × 72 in. desk with ample leg room and one drawer unit.

Recommendation 3

Place the keyboard and mouse on an articulating keyboard tray.

Outcome Expected   To allow him to use his keyboard and mouse with his wrists at, or slightly below, his elbow height, his keyboard and mouse should be placed on an articulating tray. There is a wide-open space (50 in.) where he can position the tray, and he is advised to consider the optimal area for his computer.

Recommendation 4

Speed up the "Motion" of the mouse.

Outcome Expected   To reduce his right arm and wrist motions to operate the mouse, he was shown how to increase the "speed" of the mouse in the Windows control panel. The speed for the mouse was increased from 50% to 100%, and he will modify this as he becomes accustomed to its functioning.

Recommendation 5

Learn keystroke alternatives to mouse clicks to reduce mouse use.

Outcome Expected   To reduce arm movement between his keyboard and mouse, he was shown a comprehensive list of keystrokes that replace mouse clicks. He bookmarked the 35-page list that is available at the evaluator's website.

Recommendation 6

Try using a wide document holder directly under the monitor.

Outcome Expected   To prevent him from having to look down and turn his neck to view reference documents as he works on his computer, he should purchase a wide document holder that can be positioned directly under his computer monitor.

Recommendation 7

Raise the monitor 3 in.

Outcome Expected To eliminate his neck flexion and his forward head as he views his monitor, a 2-in.-thick ream of copy paper was place between his monitor and CPU. This places his horizontal line of sight approximately 2 in. below the top of the screen and places his head and neck in neutral postures, although another inch of elevation would be optimal. He may wish to use a monitor riser to elevate it the full 3 in., but he should wait until he has used this 2-in. elevation for a few days, and when the replacement desk is installed it can be elevated about ¾ in. using by the screw-out feet.

28. GEORGE MANION: *UPPER BACK*

History

George Manion is right-handed, 6 ft 1 in. tall, and has worked as an associate professor of economics at Acme University for 15 years. He has worked at his current workstation for approximately seven years and has been experiencing discomfort in his upper back. He reports that his discomfort was caused initially by a recreational incident when he was very young, and that he later incurred a fractured disk in the thoracic spine before his work at the university. He states that he had an onset of back discomfort one month prior to the evaluation which was so severe that he went to the emergency room. Although that onset appeared to be caused initially by recreational activities, his time spent sitting at, and moving between, his desks may have contributed to it.

Observations

Mr. Manion works in an office with a large desk as a primary work surface that is 60 in. × 36 in. and 30 in. tall. He has a separate computer station, a 60 in. × 30 in. table that is 30 in. tall and has a semiadjustable keyboard surface. The keyboard does not have room for a mouse and it cannot be positioned higher than 26 in. from the floor. As a result of this configuration, Mr. Manion will elevate his right hand, arm, and shoulder 4 in. above his keyboard height to operate his mouse, which is placed on the desk and creates an awkward posture as he works on his computer.

The keyboard height is also problematic at 26 in. because his seated elbow height is 28 in. in his chair, and this places his wrists considerably lower than his seated elbow height, which is not recommended.

Additionally, since he does not type by touch, he will frequently look down at the keyboard to locate the keys, and having the keyboard so low increases the neck bend required to see the keys.

His chair appears to be an old semiadjustable task chair with cushioning that has failed over time, and its seat pan is not deep enough for his leg length. Also, the backrest is at its lowest position, which places the lumbar protrusion below his lumbar lordosis (inward curve). That protrusion is aligned with his hips and creates a gap above that at his lumbar spine, which contributes to him sitting in a forward arch, especially in his upper back, as he works at his computer.

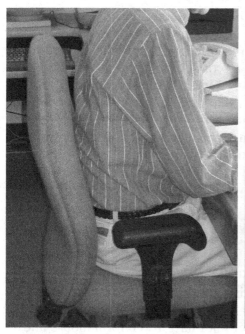 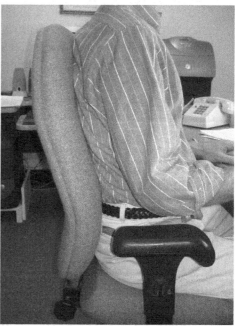

Initial backrest position was low and placed the lumbar support below his belt, creating a gap above.

With the backrest raised 3 in., the backrest curves align with his spine to provide total back support.

He reports that he spends about 50% of his day at his computer and will frequently move his chair back and forth between his computer desk and his noncomputer desk. To do this, he is observed to push, pull, and maneuver his chair by digging his heels into the carpet and grasping the work surface so that he can use his arms and shoulders to pull the chair. This is especially difficult since the wheels of this chair do not roll very easily, and the two locations are separated by approximately 6 ft of carpeted floor.

He has his monitor placed on the desk surface, which positions his horizontal line of sight above the top of the monitor cabinet and causes him to bend his neck and look down to view the lower sections of the screen. Having the chair backrest low along with this low monitor height contributes to rounded shoulders and hunching forward.

Objectives

There are four primary objectives for improving his work postures and reducing the impact to his lower back: (1) to devise a way to position his keyboard and mouse at his seated elbow height, (2) to raise his monitor and reference documents, (3) to reduce the distance and effort to move his chair between desks, and (4) to adjust or replace his current chair for improved back and upper leg support.

Recommendation 1

Replace the current table with a 27-in. tall 30 in. × 60 in. computer table.

Outcome Expected   Because his computer table places his keyboard too low and the mouse too high, the optimal and easiest solution would be to replace this table with another 30 in. × 60 in. table that is 27 in. tall. These tables have adjustable feet for additional height adjustability and are available from the university warehouse. The 27-in. height will position his wrists slightly below his seated elbow height and allow the keyboard and mouse to be at the same level. Such tables are available from the Acme used furniture warehouse.

Recommendation 2

Raise the monitor and use a document holder between it and the keyboard.

Outcome Expected   At the time of the evaluation, his monitor was elevated 2 in. by using a 1-in. monitor stacker, and an 18-in.-wide document holder, which includes another 1-in. stacker as its support. This creates a total elevation of 2 in., and he was given another 2-in. stacker, to attain a total elevation of 4 in. He should wait one week before installing the 2-in. stacker because a change of monitor and neck positioning of more than 2 in. at a time can cause discomfort.

With keyboard at 26 in., he has a significant bend in his upper back to view keyboard. Monitor and backrest are too low.

With the monitor raised and a document holder between his keyboard and monitor, he sits upright, against the raised backrest.

29. BARBARA JEAN TRENTON: *WRIST*

History

Barbara Jean Trenton is right-handed, 5 ft 8 in. tall, and has worked at Acme for eight years. She has worked in her current office for nine months and has been experiencing discomfort in her right hand. She reports that she began feeling discomfort along the center of her right palm three months ago and that it has been steadily getting worse. She has difficulty with handwriting and other tasks that require a right-hand finger grip. She notes that she felt some improvement during the holiday break, but that it has again become problematic and she has sought medical attention.

Observations

Ms. Trenton works at a height-adjustable corner-oriented work surface that she positions 29.5 in. from the floor. She has her standard keyboard and large mouse placed on the work surface and does not use a wrist rest for her mouse. As she operates her keyboard, she abducts her arms outward, placing her elbows on the desk, which flexes her shoulders. With her entire forearm on the desk as she operates her mouse, she positions the mouse back of the keyboard and leans forward to use it.

Her contoured mouse is quite large and she demonstrates 35° of wrist extension as she uses it, along with contact pressure of her wrist crease against the work surface.

She sits in a standard semiadjustable chair that she elevates to its highest position of 20.5 in. from the floor, placing her hips at her knee height. She states that she would prefer to sit higher than this, but her chair will not go higher. This chair height places her seated elbow height 27.5 in. from the floor, which is 2 in. lower than her work surface height, and contributes to the placement of her elbows on the desk surface. This is a common posture when seated elbow height is lower than desk height, and it tends to pull the person forward away from the chair's backrest. The backrest is small for her and does not provide any upper back support when it is aligned for her low back.

Her monitor is placed on the desk surface and her horizontal line of sight meets the screen at the very top edge. She has the screen angled back at the top and is observed to tilt her head down to view the lower sections of the screen, which contributes to her sitting forward of the chair backrest.

Objectives

The primary objectives for Ms. Trenton are (1) to provide her with a chair that rises higher and has upper back support, (3) to position her work surface so that her wrists are slightly below her seated elbow height (with a wrist or palm rest), and (4) to elevate her monitor so that her horizontal line of sight meets the screen about one-third of the way down from the top.

With the work surface higher than her seated elbow height, she abducts her elbows outward for typing.

With the work surface high and the monitor low, she extends her arm for the mouse and leans away from her chair.

Recommendation 1

Replace the chair with one that rises higher with a tall backrest.

Outcome Expected   At the time of the evaluation, Ms. Trenton tried sitting in some of the fully adjustable chairs that the EH&S department has for special situations. She found that the model made for tall people (5 ft 7 in. +) was the best fit and had a backrest that aligns very well with her spine, especially since it has a depth-adjustable lumbar support. This chair rises to 22 in. from the floor, which is an optimal height for her lower extremities. She will try this chair for a few days to measure the benefit. This higher seated height places her seated elbow height at 29 in. from the floor.

With the work surface above elbow height and no upper back support, she leans on her forearms.

Desk lowered, document holder below the elevated monitor, upper back support, second mouse (left).

Recommendation 2

Position the work surface at, or slightly below, seated elbow height.

Outcome Expected   To allow her to work in neutral upper extremity postures, the work surfaces were lowered approximately ½ in. to match her seated elbow height of 29 in. in the taller chair. This positioning helps her sit back in her chair with her upper arms parallel to her torso.

Recommendation 3

Use a wrist rest for the mouse.

Outcome Expected To reduce the wrist extension and contact pressure against the desk as she uses the mouse, she was given a high-quality gel wrist rest that conforms to the area of contact and distributes the weight on it evenly. This gel wrist rest can be chilled to serve as an anti-inflammatory.

35° of wrist extension and contact pressure on desk.

Wrist angle is neutral with even cushioning at wrist or palm.

Recommendation 4

Raise the monitor and move it closer.

Outcome Expected Since Ms. Trenton sits higher in the new chair and her desk was lowered, her monitor was lower than it had been initially. Therefore, she was provided with two 2-in.-tall monitor stackers that position her horizontal line of sight one-third of the way down from the top of her monitor. This positioning helps her to maintain an upright posture and to receive the upper back support that her tall-back chair provides.

Recommendation 5

Consider the use of a wide document holder below her monitor.

Outcome Expected   Since her monitor screen is now much higher than her desk surface, there is room to place a document holder between the screen and the keyboard. This will also assist her to maintain an upright posture since she will not need to look down for documents on her desk.

Recommendation 6

Leave the original mouse installed for left-hand use.

Outcome Expected   Since she uses a large contoured mouse with her right hand with which she feels comfortable, she is encouraged to install her original mouse and place it to the left of her keyboard. This will allow her to use her left hand for some nonprecision mouse activities and allow her right wrist and arm to rest.

30. HELEN FOLEY: *WRISTS*

History

Helen Foley is right-hand dominant, 5 ft 7 in. tall, and has been employed as a manager of financial analysis at Acme for three weeks. She reports that she has occasional discomfort in her wrists, and that she was diagnosed with right carpel tunnel syndrome eight months ago. One pronounced symptom is numbness in her right thumb, and she acknowledges that she often sleeps with her wrists folded.

Observations

Ms. Foley works at a wooden desk that has a 30 in. × 62 in. main work surface and a shallow right-side desk "return." The desk is 29 in. high and the return is 26 in. high. Her keyboard is placed on an articulating keyboard tray that is only wide enough for the keyboard, and she has a separate articulating platform for her mouse, positioned to the right of the keyboard tray. These are both positioned at 24 in. from the floor, attached to the return surface, and she does not use wrist rests.

Because the desk-return surface is only 22 in. deep, the monitor is positioned to the left of that surface, at an angle so she can maintain a 24-in. viewing distance. As she views the monitor at this height, she is observed to flex her neck downward, which also pulls her forward in her chair. She sits on the front edge of her chair because the recline feature cannot be locked and there is no tilt resistance at all. As a result, she must engage her torso muscles in order to sit upright at her computer, since leaning back to relax her muscles will cause the backrest to fully recline. She will also place her right foot on the chair seat pan at times. This chair is not ideal for computer work and the fixed-height armrests are very hard, narrow, and do not rise to her seated elbow height of 28 in. from the floor.

Initial configuration with nonadjustable executive chair that reclines very easily. (The tilt tension is low and the backrest does not lock.)

Right arm is significantly abducted for her mouse when viewing the monitor. A thick mouse pad was inserted to elevate the mouse and reduce wrist extension.

Recommendation 1

Replace the existing keyboard and mouse trays with a single 27-in.-wide platform keyboard tray on the main desk.

Outcome Expected   Because her keyboard tray is very low when attached to the 26-in.-high surface and it does not have room for the mouse, it should be removed. Because the mouse platform is separate and is mounted to the right a number of inches from the keyboard tray, it does not allow the mouse to be close enough to the keyboard, creating the necessity for an extended reach.

Ideally, she will have her keyboard and mouse elevated to approximately 28 in. (her seated elbow height) on a single 27-in.-wide platform. Her shoulders, arms, and wrists would then be positioned in neutral postures. An articulating tray will be angled slightly to the right, which will allow her to work with her torso square to her monitor. The monitor will be positioned toward the corner of her desk so that it does not block her view of those sitting across from her. A 27-in.-wide keyboard tray will allow her mouse to be positioned immediately adjacent to her keyboard.

Recommendation 2

Tighten the tilt tension of the current chair.

Outcome Expected   To allow her to sit back in her chair and receive some upper back support, the tilt tension was adjusted from its lowest setting to approximately 80% of the tension. This allows her to sit back and have the chair support her and will also eliminate her tendency to sit with one leg under her as she works.

Recommendation 3

Consider a more adjustable task chair.

Outcome Expected   Because the "executive style" chair she currently uses has a seat pan that is too deep for her and has no adjustments other than height and tilt tension, she would benefit from a fully adjustable task chair. During the evaluation a small semiadjustable task chair was tried, which she felt was a significant improvement. This chair must be returned to the owner, but the smaller seat pan allowed her to sit back against the backrest.

Significant wrist extension with front of tray at 24 in.　Wrists are neutral when keyboard tray is level at 28 in.

Recommendation 4

Reposition the current keyboard platform to be horizontal.

Outcome Expected To reduce her wrist extension as she types on the keyboard, the tray was moved into a more horizontal position. This is a temporary fix that should include a wrist rest until she has the keyboard tray replaced, as noted in Recommendation 1.

Recommendation 5

Increase the mouse "speed" to reduce wrist and arm movement. (This is identified as "pointer speed" in the Microsoft control panel—mouse.)

Outcome Expected Her mouse speed was increased from 50% (the default) to 100%, which she was easily able to accommodate. Her arm and wrist motions will be reduced as a result of the screen cursor responding to less movement of the mouse.

Recommendation 6

Try installing a second mouse for her left hand.

Outcome Expected   Her computer will allow a second USB mouse to be connected. That mouse should be placed to the left of her keyboard so she can operate it with her left hand. She will use the left-side mouse only for simple mouse activities that do not require precision or speed, and use her right-side mouse for work that does require precision.

31. WARREN MARCUSSON: *WRIST*

History

Warren Marcusson is right-hand dominant, 6 ft tall, and has worked as a project coordinator at Acme for one year. He has been in his current office for eight months and has been experiencing significant discomfort in his right wrist. He reports that his discomfort began about three months ago as his computer work increased, and has steadily been getting worse.

Observations

His keyboard and mouse are placed on an articulating keyboard tray that has a wrist rest for the keyboard and separate platform for the mouse that does not have a wrist rest. As Mr. Marcusson operates his mouse, he demonstrates approximately 35° of wrist extension.

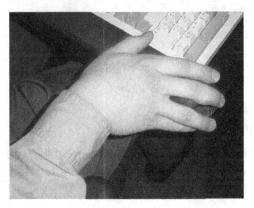

As Mr. Marcusson demonstrates his work activities, it appears that he is able to move his wrists through a wide range of motion. He was asked to place his thumb directly against the inside of his forearm, which he could do easily. Accordingly, Mr. Marcusson demonstrates hypermobility in his wrists and also in his fingers, which allows him to use exaggerated wrist angles as he operates his computer. The exaggerated wrist angles and the repetitive mouse work are likely contributors to his current discomfort.

Recommendation 1

Consider using an "upright" mouse.

Outcome Expected   At the time of the evaluation, Mr. Marcusson was asked to try a new mouse product. The new upright mouse keeps the wrist in a very neutral posture (a handshake position) and does not require wrist movement. Because the mouse movement is produced by the upper arm and shoulder, the user is advised to monitor any new strain in those areas, as those muscles become accustomed to their involvement in mousing activities. They are larger muscle groups and, typically, can absorb these motions will minimal impact.

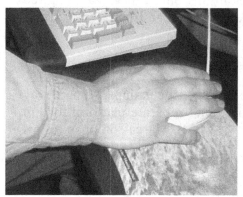

With his keyboard low, he has up to 50° of right wrist extension.

Wrist extension and ulnar deviation are reduced with tray raised and upright mouse.

Recommendation 2

Raise the keyboard tray so that his wrists are close to his seated elbow height. If they remain lower than elbow height, position the tray with a negative tilt.

Outcome Expected   His wrist extension can be reduced by raising his keyboard tray so that his wrists are at the same height as his elbows or by placing the keyboard at a slight negative tilt. The tilt will often cause the mouse to roll off the back of the mouse platform. To eliminate this problem, a trackball can be used, which also reduces wrist movements, or a small strip of self-adhesive weatherstrip can be attached to the back of the platform to contain the mouse.

Recommendation 3

Try to reduce wrist angles to less than 15° and let them rest.

Outcome Expected His extreme wrist angles, combined with more wrist activities due to his recent increase in computer work, may collectively be causing overuse and a lack of sufficient recovery time. Reducing the degree of his wrist angles in all activities, and allowing break periods of nonactivity with neutral wrist postures, will be beneficial to his muscle recovery. Some stretching diagrams were provided for him to perform conservatively.

32. HENRY BEMIS: *WRIST, ELBOW*

History

Henry Bemis is right-handed, 6 ft 1 in. tall, and has been employed as a technical support representative at Acme for seven months. He has been at his current computer workstation for the last three weeks and has a diagnosis of right upper extremity overuse. Mr. Bemis wears glasses during the workday. Currently, he works 4 days/week for 10 hours/day. His job responsibilities are to coordinate technical support functions for a single high-volume customer. His work entails significant phone work and computer input. He does very little handwriting and very seldom works from reference documents.

According to Mr. Bemis, a few months ago, he began experiencing minor irritation in his right elbow, which migrated into his wrist. His discomfort increased and began to impede his ability to operate a mouse as well as to perform other activities with his right hand. He reported that he had a right shoulder injury seven years ago as the result of falling 30 ft during a rock climb.

He is being treated by a physician and has been prescribed to receive physical therapy, a wrist splint, and work restrictions that require a break every hour.

Job Activities

Repetitive/static upper extremity motions
Sitting stationary
Static wrist postures
Reaching

Observations

Mr. Bemis works at a modular corner workstation with the work surfaces set 29 in. from the floor. His keyboard and mouse are placed on
an articulating keyboard tray that is positioned in a steep positive angle
with the front edge 24 in. from the floor. This places his keyboard just
above his upper legs and his wrists are well below elbow height as he
types.

His keyboard tray did not initially have a mousing platform, so his
mouse was placed on the work surface, which required an extended
reach to operate it. In the last week he installed a mouse platform on
the right side of the keyboard tray, which has reduced his right arm
extension. He demonstrates 30 to 40° of wrist extension as he types
and uses his mouse because of the low height of his keyboard tray and
because of the positive tilt the tray is in.

His phone must be triggered by depressing buttons on the unit,
which is positioned at the far back edge of his desk. This causes an
extended right arm reach and a right-side lean. He states that he will
move the phone closer to him, but that often creates problems with the
headset cord.

Initial posture with keyboard low and angled   Initial reach for phone; not fully achieved yet
down.   in photo.

He has his chair adjusted to 18 in. from the floor, which places his
knees slightly above his hip height. This is lower than optimal for his
height, but he reports no lower extremity discomfort and likes to have
his feet solidly on the floor. He also mentioned that his chair will not

go up any higher because the mechanism has failed. Additionally, he sits this low so that he can slide his keyboard tray under his desk without lowering the mechanism. The sharp keyboard angle lowers the clearance height for his legs.

He demonstrates a typing style in which he uses a light touch to depress the keys and will often elevate his wrists to reduce wrist extension. When using his mouse, he places his right wrist on the mousing platform and has approximately 40° of wrist extension as he operates it. He states that he may fold his wrists as he sleeps, which can contribute to his discomfort. He is also an avid player of computer games, using the typical small game controllers.

Recommendation 1

Adjust the keyboard tray to a more horizontal position.

Outcome Expected At the time of the evaluation, Mr. Bemis was shown how the angle of his keyboard tray can be adjusted. It was placed in a more horizontal position to reduce his wrist extension and elbow extension. This raised the front edge of the tray, which allows more leg room, and his chair can be raised higher.

Initial keyboard position, low and angled.   Keyboard tray raised and placed more horizontally.

Recommendation 2

Try using a trackball to reduce wrist motions.

Outcome Expected At the time of the evaluation, a co-worker's trackball was installed to reduce wrist motions. Mr. Bemis was readily adept at using this style of mouse device and should mention this modification to his physical therapist. He still has wrist extension with it.

Recommendation 3

Try using a wrist rest with the trackball.

Outcome Expected   At the time of the evaluation, he was allowed to try a gel wrist rest that is very malleable to distribute the weight on it evenly. This reduces his wrist extension significantly, and he will rest his thenar pad on it as well as his wrist, to reduce wrist compression.

Initial trackball posture with wrist extension.   Five-inch gel wrist rest positioned to reduce extension.

Recommendation 4

Adjust his chair height, seat pan depth, and arm rest height.

Outcome Expected   The evaluator demonstrated how this model chair has an unusual feature for raising the seat pan height. The right-side lever is pressed downward to allow the seat pan to rise upward and is quite difficult to control. The seat pan was elevated slightly, since the front edge of the keyboard tray is higher after the angle change, providing more leg clearance. He was also shown how his chair has an adjustment for the depth of the seat pan. He extended the seat pan forward for more upper leg support.

Recommendation 5

Leave the mouse installed to operate with his left hand.

Outcome Expected   To allow recovery time for his right wrist, he is advised to do some mousing with his left hand. He mentions that he will occasionally reach his left hand across his torso to operate his mouse while it is on the right-side mousing platform. Therefore, he is able to operate a mouse with his left hand in the computer applications he performs, and will be able to acclimate to this easily. His computer can accommodate multiple mice, and his keyboard tray can accept a mouse platform on either side. He can obtain a second mouse platform internally and he was advised that the mouse buttons must be configured identically when using more than one mouse.

33. AL DENTON: *WRIST, ELBOW, FOREARM*

History

Al Denton is right-hand dominant, 6 ft 3 in. tall, and has worked as a software engineer at Acme for two years. He has been in his current office for one and a half years and has recently been diagnosed with tendonitis in his left elbow. He reports that his initial symptoms developed about a year ago and were exacerbated by a project that required long hours at his computer. He also reports moderate discomfort in both wrists and forearms. Mr. Denton wears glasses during his workday and works a 40-hour workweek.

Job Activities

Sitting
Computer input activities: mousing and keyboarding
Minimal handwriting and phone work

Observations

Mr. Denton works at a modular corner workstation that has the surfaces set at 33 in. from the floor, with his keyboards and mice placed directly on the work surface. He works with two computers, and each has an angled keyboard and a wheel mouse. He does not use wrist rests and demonstrates approximately 30° of wrist extension (bending back) as he operates his keyboard and mouse. He places both keyboards and mice approximately 10 in. back from the front edge of the desk and rests his forearms on the desk surface in front of them.

He sits in a semiadjustable chair that he positions 20.5 in. from the floor. This seat pan height places his hips at the same height as his knees, which is optimal. His main computer is placed in the left corner of his desk, and his Windows computer is positioned on the right-side work surface. He often moves back and forth between them and will frequently have one hand on each keyboard as he reviews the corresponding data. When he works on the Windows computer he cannot move his legs completely under the keyboard because he has two CPU towers placed below the monitor. This causes him to adduct his left arm across his torso to reach that keyboard. Similarly, he will also adduct his right arm across his torso to the left for work on his main computer's keyboard.

With his seat pan height at 20.5 in., his seated elbow height is 30 in., which almost 12 in. higher than his chair height. This is an atypical variance and is partially accommodated by the length of his upper torso. However, with his work surfaces set 33 in. from the floor, his wrists are considerably higher than his elbow height as he operates his computers, which is not advised.

With his seated elbow height 30 in. from the floor and his work surfaces set 33 in. from the floor, he must elevate and abduct his shoulders
and arms to place his elbows on the work surface. This posture often
has his left forearm firmly against the nonrounded desk edge just below
(toward his wrist) the location of his forearm discomfort. This contact
pressure can be damaging when the posture is prolonged.

As he performs his computer tasks, he demonstrates a very frequent
action with his left hand, where two computer keys are pressed simultaneously. One version of this motion places his left little finger on the
left "Ctrl" key as he presses other left-side keys with other fingers. He
calls this motion his "pinky-pivot" and performs it very fast with significant force. He use more force than needed for all of his keyboard
work, and his keystrokes become very audible.

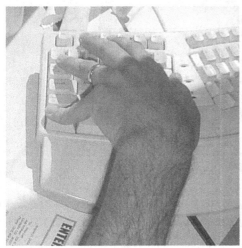

Ulnar deviation for "pinky pivot."

Ulnar deviation when using any style of
keyboard.

His horizontal line-of-sight meets the monitor screen about mid-screen (vertically) and is observed to tilt his head back as he views the middle and upper sections of his monitor.

Objectives

The primary postures and actions that appear to contribute to his symptoms are the ulnar deviation of his left wrist as he rapidly performs combination keystrokes, and the contact pressure of his forearms on the desk surface/edge. Additionally, the force he uses for his keyboard keystrokes becomes a stressor to his hands and arms, as well as the reach across his torso for the keyboards that are separated by almost 3 ft. The height of his desk contributes to his discomfort, since his wrists are higher than his seated elbow height.

Recommendation 1

Try using a cushion along the front edge of the desk.

Outcome Expected To reduce the contact pressure of the desk edge against his forearms, he was given a 36-in.-long cushion that is designed for the front edge of a work surface. It is long enough to be positioned in front of both of his keyboards (for the left arm), and he will try to find a way to secure it in place.

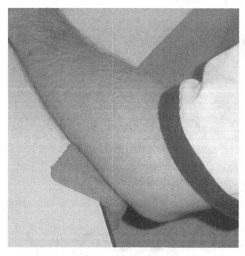 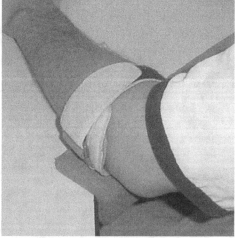

Desk edge cushion without elbow brace.

The cushion makes wearing the brace comfortable.

Recommendation 2

It would be advisable to use less force when pressing the keyboard keys.

Outcome Expected To lessen the muscle impact from striking the keyboard keys, he should try to use less force as he types. This is a very hard habit to break, and he will try to post a reminder note to prevent him from hitting the keys so hard. He should also try to avoid the extreme left wrist angles.

Recommendation 3

Lower the work surface or use keyboard trays.

Outcome Expected To position his wrists at, or slightly below, his seated elbow height, he should have his work surfaces lowered or install keyboard trays for all of his computer devices. Lowering the work surfaces to 30 to 31 in. is the preferred modification, so that his monitors will be positioned lower and reduce his neck extension as he views the upper sections.

Recommendation 4

Lower the monitors if the work surfaces are not lowered.

Outcome Expected If he does not lower his work surfaces, he should lower his monitors so that his horizontal line of sight meets the screen one-third of the way down from the top.

Recommendation 5

Move one of the CPU towers so that he can move farther to the right.

Outcome Expected To reduce his torso lean and left arm adduction to use his right-side keyboard, he should move one of the CPUs out of the right-side under-desk area so that he can move his legs fully under the right-side keyboard.

34. MARSHA WHITE: *WRIST, ELBOW, FOREARM*

History

Marsha White is right-handed, 5 ft 2 in. tall, with a diagnosis of right lateral epicondylitis. According to Ms. White, she began experiencing right lateral elbow and forearm pain early in August that migrated into her right wrist along the lateral side. These episodes of discomfort would manifest as pain in the upper lateral aspect of her right forearm. When her recent symptoms had grown into numbness from her right shoulder to her fingertips, she decided to seek medical attention. She serves as the head of circulation services and has been in her current office since beginning her employment. Ms. White wears glasses during the workday. Currently, she works 5 days/week for 8 hours/day. She is receiving two sessions a week of physical therapy and wears a rigid wrist splint on her right wrist during the workday.

Job Activities

Sitting stationary
Static wrist postures
Reaching outward and above the shoulders

Observations

She was asked to remove her wrist splint to observe her wrist postures when using her computer. She is a fast typist and demonstrated ulnar deviation in both wrists up to 30°. This ulnar deviation is exacerbated by the fingers she uses for the tab and backspace keys. As shown in the photos, she uses digit 3 in each hand to press those keys. She also uses an atypical tight pinch posture for handwriting with a standard shaft pen.

Ms. White's main work areas are her desk and her computer workstation. Her desk is a 30 in. × 60 in. work surface 29.5 in. from the floor, with a left-side return 26.5 in. from the floor. She has her phone on the return surface and does handwriting on both surfaces. Her desk chair is a nonadjustable side chair that has armrests but no wheels. When she works between the two desk surfaces, with a chair that does not swivel, she will twist her upper torso to reach the work surface not aligned with her chair's orientation, or she must lift her chair by the armrests

Left ulnar deviation.

Right ulnar deviation.

to reposition it. These chair movements are awkward and require significant effort on an ongoing basis.

Recommendation 1

Consider using an angled keyboard.

Outcome Expected Because she ulnar-deviates her wrists as she types, a split or angled keyboard will be a solution to help her maintain a neutral wrist posture when typing.

Recommendation 2

Consider using her left hand for mouse operation and/or an upright mouse.

Outcome Expected Because she experiences significant discomfort from mouse clicking with her right index finger, she may find relief from mousing with her left hand and/or by trying an upright mouse for her right hand. The upright mouse positions the hand in the most neutral posture possible, and clicking is performed on an upright mouse by rocking the right thumb to the left or to the right on the paddle switch at the top of the vertical handle. Her computer will support two mice simultaneously.

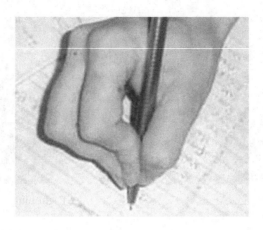

Recommendation 3

Use a pen that has a wider, cushioned shaft. (Many such pens are now available.)

Outcome Expected At the time of the evaluation, a 2-in. section of foam tubing was placed over the shaft of her pen to reduce her tight pinch and to provide a more tactile surface that will allow her to grip the pen with less force. That modification is shown in photos of her keyboard with the pen resting on it. Her initial posture is shown here.

35. GRACE STOCKTON: *WRIST, ELBOW, FOREARM, SHOULDER*

History

Grace Stockton is right-hand dominant, 5 ft tall, and works as an accounts payable specialist. Ms. Stockton has been at Acme for six months and has been at her current computer workstation since beginning her employment. She has a diagnosis of bilateral arm/shoulder pain R > L.

According to Ms. Stockton, she began experiencing discomfort across her dorsal right wrist in September that would manifest as pins and needles, numbness, and tingling. Those sensations migrated into her right forearm, elbow, and shoulder during the next few weeks.

Observations

Ms. Stockton works at a modular corner-oriented workstation with the work surfaces set 29 in. from the floor. She reports that her initial posture had been to sit on the from edge of her chair with her angled keyboard and mouse on a keyboard tray that was very low, causing her to reach down for her keyboard and mouse on the tray with her wrists bent back. She put it into the lowest position. She felt that positioning was uncomfortable, so she recently moved her keyboard onto the desk surface and pushed the keyboard tray back under the desk, out of the way. She also uses a large 10-key calculator positioned to her right.

Having her keyboard and mouse on the desk surface allows her to work with her wrists in a more neutral posture; however, she must elevate her semiadjustable task chair to its highest position of 20 in. from the floor, which prevents her from being able to rest her feet firmly on the floor. She will typically place her feet on the chair legs and/or sit on the front edge of the chair and extend her arms forward for her keyboard and mouse, which eliminates any support from her backrest. Sitting at this height places her seated elbow height 28 in. from the floor, which is 1 in. lower than the work surface height. Accordingly, to type and use her mouse, she extends her arms forward to elevate her elbows to her wrist height. However, this moves her elbows forward, rounds her shoulders, and contributes to sitting forward with no support from her chair.

Recommendation 1

Move the chair's seat pan lower so that her feet rest firmly on the floor and align the lumbar support.

Outcome Expected At the time of the evaluation, her chair was lowered 2 in., which allowed Ms. Stockton to sit back in the chair and have her feet firmly on the floor. This provided support for her back, including the upper back. The lumbar support was lowered approximately 3 in. to align with her lumbar lordosis (inward curve).

Recommendation 2

Replace the angled keyboard with a conventional keyboard and replace the keyboard tray wrist rest.

Outcome Expected At the time of the evaluation, her previous keyboard was acquired and reinstalled, along with the wrist rest designed for her keyboard tray. This reduced her wrist extension and bought her hands inward because the keyboard is not as wide. She reported that it felt much more comfortable.

Recommendation 3

Adjust the keyboard tray into a slight negative tilt.

Outcome Expected At the time of the evaluation, her keyboard tray was angled in a slight negative tilt, which placed her wrists in a very neutral posture and allowed her to move her chair in closer to her keyboard tray because the chair armrests move under the front of the tray.

Recommendation 4

Try using a wide in-line document holder.

Outcome Expected At the time of the evaluation, we reviewed the benefit of an 18-in.-wide document holder positioned between her keyboard and her monitor. Using it prevented her from leaning forward over her keyboard to view the invoices and promotes a neutral upper extremity posture. This modification was not photographed.

Elbows forward and shoulder flexed with her keyboard and mouse on the desk surface.	With the keyboard and mouse on the tray, at elbow height, arms and shoulders are neutral.

Recommendation 5

Angle the monitor downward and move her desk tools closer.

Outcome Expected Because her chair was lowered 2 in., her monitor became relatively higher. The higher monitor height appears to be better, as it induces her to sit back; however, she felt the change was somewhat awkward. To compensate initially, the monitor angle was changed from being angled upward to being angled slightly downward. This position felt more comfortable for her and also moves the reflection of the overhead light fixtures outside her field of vision. The screen glare that had been troublesome to her is no longer in evidence.

36. SAM CONRAD: *WRIST, FOREARM, LOWER BACK*

History

Sam Conrad is right-handed, 6 ft tall, and has worked as a customer services representative at Acme for five years. He has been in his current work area for approximately six months and has recently experienced a significant onset of recurring lower back pain. Additionally, he notes that he has discomfort in wrists and forearms R > L since moving to this new work area.

He reports that he has had lower back discomfort since an incident 10 years ago and had a discectomy two years ago. He has had significant improvement since that surgery; however, he states that he began to experience a recurrence of his lower back pain in March of this year. He is very disciplined about performing his daily home exercise program, but cannot sit for extended periods of time.

Job Activities

Sitting
Computer input activities: mousing and keyboarding
Repeated phone conversations
Occasional handwriting

Observations

Mr. Conrad works at a modular corner workstation that has independent crank-adjustable work surfaces for his keyboard/mouse and monitor. The front section that holds his keyboard and mouse is positioned 28.5 in. from the floor, and the monitor section is positioned at 31.5 in. from the floor.

Mr. Conrad sits in a semiadjustable chair that he has positioned at 22 in. from the floor with the armrests at 32.5 in. from the floor. The seat pan height places his lower extremities in neutral postures and the armrests accommodate his short upper arm length; however, they are approximately 4 in. higher than his keyboard and mouse. Having the armrests so much higher than his wrists as he types and operates his mouse creates significant contact pressure against an isolated area at the underside of his forearms near his elbow. He reports that his forearm discomfort is at that location in his arms.

As he views his monitor, his horizontal line of sight meets the monitor at the very top of the cabinet. As a result, he will flex his neck and upper spine forward and lower his head in a slouching posture, which also rounds his lumbar spine.

He has experimented with all the adjustments on his chair, especially the backrest. He has the backrest in a vertical position and at its highest position; however, the lumbar protrusion is approximately 2 to 3 in. lower than his lumbar lordosis (inward curve) and it is such a narrow protrusion that it provides support only to a very isolated area of his spine.

There is a large chair nearby which Mr. Conrad states is more comfortable for him. He and the evaluator sat in this chair and determined that the increased comfort arises from the lumbar protrusion being higher and covering a larger area of the lumbar spine. However, the chair's armrests are set too far apart for him.

Forward lean with head down for low monitor; rounds the spine.

Monitor raised and back cushion for broader support to low back.

Recommendation 1

Install an inflatable lumbar cushion to raise and disperse lower back support.

Outcome Expected   At the time of the evaluation, he was given an inflatable lumbar cushion which was lightly inflated and positioned so that it provides support slightly above the chair's lumbar protrusion and spreads the support through a larger area. However, this cushioning does prevent him from receiving support to his upper back, and another chair with a less pronounced lumbar support would be preferable.

Recommendation 2

Raise the monitor platform 3 in.

Outcome Expected   At the time of the evaluation his monitor riser was elevated approximately 3 in. so that his horizontal line of sight meets the screen 2 to 3 in. below the top of the screen. This has the effect of moving his upper torso back and into a more upright posture.

Recommendation 3

Raise the keyboard and mouse surface and lower the armrests.

Outcome Expected To help him sit back, the keyboard and mouse platform was raised about 1 in. so that his wrists are slightly higher. To allow him to move his upper arms back, his armrests were lowered 2 to 31.5 in., which also reduced the contact pressure (and discomfort) against his forearms.

Recommendation 4

Speed up the "motion" or "acceleration speed" of the mouse.

Outcome Expected Because he demonstrates that he must move his right arm extensively to direct his mouse and cursor across his computer screen, he was shown how to speed up that cursor motion. This was done in the "settings–control panel–mouse–motion" window. The speed was raised from 50% to 100%, which allows him to move the mouse through a shorter distance to achieve the same cursor distance across the screen.

Recommendation 5

Use a footrest.

Outcome Expected At the time of the evaluation, a footrest was placed under his feet to move his posture to a more upright position. He found this very beneficial, and he may alternate between using it and not using it to allow him to sit for longer periods of time without an onset of lower back discomfort.

Recommendation 6

Use a gel mouse rest for his right wrist.

Outcome Expected To reduce his wrist extension when using his mouse, as well as the contact pressure against his right wrist that he rests on the work surface, a gel wrist rest was placed in front of his mouse pad. This reduced his right wrist extension from 40° to 20° and provided cushioning against the underside of his right wrist. The gel wrist rest used is very malleable and designed to distribute evenly the weight placed on it.

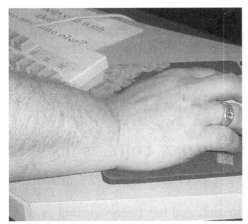 

40° of wrist extension and contact pressure on wrist.

With 5-in. gel wrist rest, wrist extension is 20° and cushioned.

Recommendation 7

Consider using an angled keyboard.

Outcome Expected Mr. Conrad demonstrates approximately 20° of ulnar deviation (bending outward) in his wrists when typing on his standard keyboard. At the time of the evaluation, an angled keyboard was installed just to demonstrate how it will reduce this wrist angle. Since he is not having wrist discomfort, this is not a modification that he should make immediately, especially since the angled keys create a learning curve that can impede productivity for a few days.

Ulnar deviation (bending outward) of 20° as he types.

Angled keyboard eliminates ulnar deviation completely.

Mr. Conrad noted that he has been performing his daily stretching routines very rapidly and demonstrates that one turning motion caused a twist in his lower back. He has since ceased performing these so rapidly, and that twisting motion may have been a significant contributor to his recent onset.

37. MARGE MOORE: *WRIST, SHOULDER, LOWER BACK*

History

Marge Moore is right-handed, 5 ft 5 in. tall, and has been working as a client services representative at Acme for one year. She has been in her current work area for 10 months and has been experiencing discomfort in her lower back, right arm, and right shoulder. She reports that she was involved in an automobile accident about five months ago, which has had a significantly negative impact on her ability to sleep restfully, engage in recreational activities, and sit comfortably at her desk. She is receiving regular chiropractic treatment for her discomfort and also notes discomfort along the top of her wrists into her forearms as she works.

Job Activities

Sitting
Computer input activities: mousing and keyboarding
Repeated phone conversations
Handwriting

Observations

Ms. Moore works at a modular corner workstation that has independent crank-adjustable work surfaces keyboard/mouse and monitor. The front section that holds her keyboard and mouse is positioned 28 in. from the floor. This height places her wrists slightly higher than her seated elbow height, which is not recommended, and the monitor is quite high, with her horizontal line of sight meeting the screen below the midpoint.

Her initial posture in her chair was to have the backrest at its lowest position, with the seat pan positioned in a rear tilt that elevated the front edge of the chair, and even at its lowest position, prevented her from resting her feet firmly on the floor. The low positioning of her backrest did not provide support to her lower, middle or upper back. The low lumbar support pushes her hips forward so that the backrest, above that point, is farther back from her torso.

She does not have a wrist rest for her mouse and demonstrates significant wrist extension as she uses it. This is caused, largely, because her corner workstation is not connected to the side surfaces and "migrates" back over time, creating a reach for her arms and hands. Her phone is positioned on the left and she is observed to adduct her right arm across her torso to operate the phone, which aggravates her right-side discomfort.

Recommendation 1

Adjust the chair's backrest height and seat pan angle.

Outcome Expected   To allow her feet to rest on the floor and provide a more upright position as she works, the angle of her chair's seat pan was moved from being angled rearward to a horizontal position. Then the backrest was raised so that the lumbar protrusion is aligned with her lumbar lordosis (inward curve).

Recommendation 2

Work with a footrest.

Outcome Expected   To further elevate her upper legs off the front edge of her chair's seat pan and provide an alternate posture for her lower extremities, she would benefit from having a footrest at her feet. At the time of the evaluation, a co-worker's footrest was tried, which was very beneficial for her.

Recommendation 3

Pull the corner surfaces out from the wall to meet the side surfaces.

Backrest raised, and she could add a folded towel over the top for increased upper back support; work surfaces close.

A footrest helps her sit with her upper back against the chair backrest and less pressure to the underside of her legs.

Outcome Expected   To create a contiguous work surface between her corner and side tables, her computer table was pulled out from the wall so that the sides meet the tables on her left and right.

Recommendation 4

Adjust all surfaces so that the keyboard surface and side surfaces are at the same height and are close together, to create a contiguous work surface with wrists at elbow height.

Outcome Expected   At the time of the evaluation, her main work surface for her keyboard was lowered from 28 in. to 27 in. so that her wrists are at her seated elbow height as she operates her computer. The side surfaces were adjusted by their screw-out feet so that they are at the exact height as the keyboard surface.

Recommendation 5

Replace the existing wrist rest with gel versions for the keyboard and mouse.

Outcome Expected To reduce her right wrist extension as she operates her mouse, she should have a small gel wrist rest positioned under it. Because her keyboard wrist rest is very firm and not a cushioned type rest, it should also be replaced with a gel wrist rest to reduce the contact pressure of her wrists against it. She is encouraged to rest the thenar and hypothenar pads of her hands on the wrist rest rather than at her wrist crease. Any pressure against the wrist crease can be damaging for extended computer work.

Without a wrist rest she has significant right wrist extension and contact pressure when using her mouse.

Right wrist is placed in a neutral position with a wallet (wrist rest) temporarily put under her right wrist.

Phone on left creates a torso twist as she operates it.

Phone on right allows her to keep her back straight.

Recommendation 6

Lower the monitor so that her horizontal line of sight is in the upper half.

Outcome Expected   To discourage her from tilting her head back as she views the upper areas of her monitor, the rear section of her adjustable computer table was lowered approximately 1.5 in.

38. JANE WILLIAMS: *WRIST, UPPER BACK, NECK*

History

Jane Williams is right-hand dominant, 5 ft 2 in. tall, and has worked as a support specialist at Acme for one year. She reports that she has been experiencing discomfort in her upper back, to the left of her right scapula and into her neck, which becomes very painful and will significantly restrict her neck movements after it sets in. She also reports wrist discomfort that she has experienced for many years, and that she occasionally works 12-hour days. She also notes that she still experiences some numbness in her hands at night and that she will occasionally wear her wrist splints as she sleeps to prevent such numbness.

Observations

Ms. Williams works at a rectangular desk workstation that is 30 in. tall with her angled keyboard and optical mouse placed on the work surface, and she does not use wrist rests.

Elbow is elevated/adducted to reach the desk height.

Elbow forward of torso with forward lean away from chair.

With her elbows lower than her wrists as she sits at her 30-in.-high tall desk, she will necessarily have her wrists and forearms in contact with the edge of the desk. The desk edge is not rounded and the wrist compression that results from the isolated area of contact pressure is a likely contributor to her wrist discomfort.

Her monitors are each elevated by two 2-in.-thick reams of copy paper, which is done because the employee who works at this station at the next shift is reportedly over 6 ft tall. This elevated monitor position places her horizontal line of sight at the middle of the screen. As a result, when she views the upper sections of the screen, she demonstrates moderate neck extension (bending back), which may contribute to her neck and shoulder discomfort.

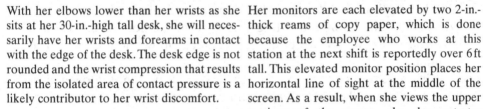

Ms. Williams sits in an adjustable chair that she positions 19.5 in. from the floor. This places her hips slightly above her knees and allows her to rest her feet firmly on the floor. She states that she has occasionally borrowed a footrest from a co-worker and that it helps mitigate her knee discomfort after sitting for prolonged periods.

The chair's seat pan is too deep for her leg length, and because the front edge of the seat pan is in contact with her lower legs as she sits back, it contributes to her tendency to sit forward, away from the backrest. The backrest is adjustable; however, the backrest angles away from the torso at the top and thereby does not offer support when she is sitting upright in the posture needed to perform computer work. She demonstrates a tendency to move her hips forward on the chair seat pan, to lower her scapulae to rest against the backrest. This is commonly done to receive temporary firm upper back support.

Objectives

The primary physical stressors to address are (1) working on her keyboard and mouse with her wrists higher than her seated elbow height, (2) extending her right elbow outward to raise her right arm to her mouse, (3) tilting her head back to view her monitor, and (4) having no support for her upper back as she works. There are also some secondary modifications included in the recommendations noted below.

Recommendation 1

Install an articulating keyboard tray at the desk.

Outcome Expected To allow her to work with her wrists at, or slightly below, her seated elbow height, she was taken to the Ergo Demo Cube in the human resources department to try an articulating keyboard tray. With her keyboard and mouse placed on an articulating keyboard tray 27.5 in. from the floor, and sitting 19.5 in. from the floor as she does at her desk, her wrists were at her seated elbow height, which is optimal.

Using a keyboard tray so that her wrists will be slightly lower than her seated elbow height will also remove the contact pressure she has of her wrists against the front edge of the desk surface.

Work surface at 30 in. places her hands and wrists 3 in. higher than her seated elbow height of 27 in.

Keyboard at 27.5 in. keeps her upper arms parallel to her torso and her wrists at elbow height. The tall backrest supports the shoulders.

Recommendation 2

Replace her chair with one that has a tall backrest that does not angle away from the torso at the top.

Outcome Expected To allow her to sit with support to her upper back while she is in an upright posture, as well as having lower back support, she was asked to try the demo chairs that are in the Ergo Demo Cube. A tall-back burgundy chair was an optimal fit for her. Before trying the demo chairs, her current chair was adjusted so that the backrest was elevated and aligned with her spine, yet she does not receive adequate upper back support from it unless she is leaning back in a posture that is not a realistic working posture.

Recommendation 3

Use a footrest.

Outcome Expected At the time of the evaluation, she tried a footrest in the Ergo Demo Cube which will provide an alternate foot position to assist her knees and to help her sit back against her backrest.

6

LONG-DISTANCE EVALUATIONS

This book is being published at a time when ergonomics is finding increasing attention around the world. Accordingly, many multinational companies are faced with the responsibility to provide ergonomic interventions to their employees in other countries—countries that often do not have ergonomists or anyone with the assessment skills to address ergonomic issues.

The solution that is rapidly evolving, with great success because of technology, is the long-distance evaluation. In this model of ergonomic evaluation, a person being evaluated is located in another part of the country of the evaluator or even in another country. I have recently conducted successful ergonomic evaluations in England, Ireland, Australia, Venezuela, Brazil, and Canada without leaving my Boulder, Colorado office.

The combination of experience with ergonomic evaluations, and the ubiquitous proliferation of digital cameras around the world, has opened the door to providing ergonomic services at very low cost to areas where access to a professional ergonomist is nonexistent. Many large companies have facilities in many countries and, more and more, have many employees who work from home. In the past, the only ergonomic resources available to many of them have been in the form of online training and printed matter.

Safety Manager's Guide to Office Ergonomics, By Craig Chasen
Copyright © 2009 John Wiley & Sons, Inc.

A long-distance evaluation begins with the evaluator being provided with the e-mail address of the employee and his or her specific location, so that the respective time zones can be factored in as well as any language barriers that may exist. With that information in hand, the long-distance ergonomist sends a note to the employee and/or the person who requested the evaluation, asking for the following information about the evaluee:

A. A description of the discomfort and three basic questions:
   1. What is the discomfort you are experiencing?
   2. How tall are you?
   3. Which is your dominant hand?
   4. How much of your workday is spent at your computer workstation?
B. Three basic measurements:
   1. The height from the floor to the surface on which your keyboard and mouse are located.
   2. The height from the floor to the top of the seat pan of your current chair.
   3. The height from the floor to the bottom of your elbows with your hands in your lap and your shoulders relaxed.
C. Photos of the employee in typical working postures. I ask the person to incorporate the area(s) of discomfort in most of the photos, but also some wide shots that show the entire chair, desk, and monitor in one image.

The way to make a long-distance evaluation effective is to:

1. *Be fluent when dealing with various file formats for photos and the way they are conveyed.* The people who provide the images may be using a wide variety of photo programs, perhaps having come with their cameras or photos provided by a photo website. Typically, the photos will be transmitted as common JPEG (Joint Photographic Experts Group) files, but even when the evaluator is using photo files taken personally, the report appearance will be enhanced by being familiar with how to manipulate digital images. It is very common to want only part of a photo you took for inclusion in an evaluation report, to highlight a specific issue or solution. Knowing how to crop photos allows great flexibility in how convincingly your findings are conveyed, by spotlighting the focus area.

2. *Try to locate vendors that can deliver products and/or services in the employee's local area.* This is a significant challenge, especially overseas. Often, the large office supply and furniture vendors have websites that indicate their presence in other countries, but the person being evaluated often has the best track on local suppliers. In the United States, a majority of suppliers have a national presence, and an increasing number of equipment and furniture suppliers have created an extended delivery and installation network to service home offices in remote areas. This is in response to the growing trend of employees working from home offices.

3. *Make sure that your instructions are simple and clear.* One thing that I have learned, through almost comical experiences, is that you really need to stress and repeat that you want the photos to show the employee being evaluated in his or her typical working posture. I have been amazed at how many times I have received photos that showed an employee's workstation very well, from a few different angles, but no one is sitting there! I have even received photos of an employee sitting in a chair with his or her back to the keyboard and monitor, legs crossed, as sitting and smiling into the camera with arms folded across the chest.

The long-distance evaluation requires practice, of course, like anything else, but as we see the increased trend toward employees working from home and also recognize the potential liability issues involved in visiting an employee's home for an evaluation, this model is bound to expand. Interestingly, as this book is written, the Occupational Safety and Health Administration (OSHA) position on "office work activities in a home-based worksite" was reversed from their initial advisory, which could have extended workplace safety regulations to people who work from their homes. They now state that "OSHA will not conduct inspections of employees' home-based offices and OSHA will not hold employers liable for employees' home offices, and OSHA does not expect employers to inspect home offices of their employees." OSHA does distinguish home offices from other home-based worksites, such as home-based manufacturing.

Continuing wide fluctuations in gas costs will surely spawn an increase in employees working from home and thereby expand the need for ergonomic interventions.

7

EXCEPTIONS AND ARCANE SITUATIONS

Any set of guidelines, including those laid down in this book, will be subject to exceptions and oddities that arise in the midst of the process. From my extensive experience in office ergonomics, the situations below represent the most common of the uncommon occurrences that arise. I offer solutions and tactics that, consistent with my theme throughout, can simplify your resolution of atypical discomfort issues. The final examples demonstrate the benefit of scouring all available data during the evaluation, to uncover causes for the discomfort reported that were not readily apparent.

Employees Who Are Especially Tall or Large

Most chairs are intended, or the manufacturers say that they are intended to accommodate the range of body sizes from the 5th percentile female (smallest frame) to the 95th percentile male (largest frame). Although the principles of ergonomics refute such "one size fits all" suggestions, the extremes at either end necessarily exist, and often create great urgency. In any midsized or larger organization, I recommend having an unused chair available that is very wide and/or without armrests. Having such a chair will be an immediate short-term solution when a new employee is hired who is too large to fit into your standard

Safety Manager's Guide to Office Ergonomics, By Craig Chasen
Copyright © 2009 John Wiley & Sons, Inc.

chairs. When a new employee arrives and he or she must wedge between the armrests of a standard size chair, it can cause embarrassment initially, and significant discomfort if it is prolonged. Therefore, it is very foresightful to have a chair on hand that can accommodate a larger person since it is not very easy to acquire a larger-than-normal chair on short notice. Even if the chair, or any solution you offer, is not perfect, providing it quickly reflects favorably on your ergonomics program.

Employees Who Cannot Sit for Prolonged Periods

With lower back problems so prevalent in today's workforce, it is increasingly common to have employees who cannot sit for prolonged or even moderate periods of time without experiencing low back discomfort. Working on a computer in a standing posture is becoming an increasingly common need for employees.

Efforts to address this situation by creating a sit/stand workstation using the keyboard trays called sit/stand trays, are usually counterproductive for employees taller than 5 ft 4 in. A sit/stand keyboard tray will very seldom rise more than 7 in. above the desk surface, putting the platform at ±36 in. from the floor at best, since most desks are about 29 in. tall. The typical standing elbow height for males is about 42 to 44 in., which places the keyboard and mouse 6 to 8 in. below their elbows. This requires them to reach down for the keyboard and mouse, causing rounded shoulders, significant wrist extension, and/or a forward lean as they work. Although an employee with lower back discomfort may be willing to make this discomfort compromise, such postures remain detrimental. Still, the limited range of typical sit/stand keyboard trays can allow a shorter person to use one and attain a neutral arm and wrist posture (wrists at standing elbow height), but a significant challenge is posed when accessing other items on the desk.

Most important is the position of the monitor in a situation where the user alternates between sitting and standing. Looking down from a standing position with the monitor low creates a very awkward neck and upper torso posture, and angling the screen does not really compensate for having the monitor at eye level.

Even with the flexibility of positioning offered by flat-panel monitors and articulating monitor arms, these solutions are effective only for computer work. On a standard-height desk, access to common items such as pens, a phone, or even documents can be very problematic from a standing position. The movements required to reach down and work with desk-level items may actually aggravate the very reasons that the standing posture is being engaged.

For the expense, effort, and limited benefit involved in the purchase and installation of such keyboard trays and monitor arms, the purchase of an electric sit/stand workstation may be a more prudent solution. The primary benefit of an electric table is that the entire desk surface raises and lowers, and because users merely push a button to change the overall height level, they are more inclined to change the surface height on an electric desk than they are when they have to adjust a keyboard tray and monitor arm manually.

Employees With Bifocals and Progressive Lens Glasses

With our aging workforce, more and more employees wear bifocals or progressive lens glasses. These multiple-lens glasses often cause neck problems because the narrow individual viewing sections cause wearers to move continually and then poise their head to attain a specific viewing position for that limited viewing area. This situation is often exacerbated when the lens sections are made even smaller by having a narrow, "more fashionable" overall lens height.

The most common section of such a lens for viewing a computer screen is at the bottom of the lens, so the monitor must be positioned very low. The frequent dilemma is that when sitting on a properly positioned desk surface, a monitor does not position the screen low enough for a neutral head posture when viewing the screen through the lower lens. Accordingly, users tilt their head back, causing neck strain, or they try to raise the chair and/or lower the desk in a fashion that causes discomfort in other areas.

The first two solutions involve removing the monitor base. With a CRT monitor, it is not often recognized that the swivel base can be removed, thereby lowering the screen about 2 in. With the base removed, they typically sit with the screen more vertical than desired (especially since it is low), but inserting a pad or two of "sticky notes" (or any means to raise the front slightly) will often raise the screen enough to improve the viewing angle and still have it much lower than with the base installed.

The second solution is to remove the base of a flat-panel monitor. Almost every flat-panel monitor has mounting holes on the back that comply with the Video Electronics Standards Association (VESA) mounting standard. VESA has developed specific guidelines for the mounting-hole pattern placement and screw size on flat-panel monitors. That standard requires that the mounting holes for monitors weighing up to 30.8 lb be 75 or 100 mm apart, and many monitors have both sets. This is so that the base can be removed and the monitor can be mounted on an articulating arm, which will allow the screen to be lowered right down to the desk surface, with only the monitor frame keeping it higher than the desk.

A third solution, used for an employee working in a modular cube with a corner-oriented computer surface, is to lower the corner section only and attach a sit/stand keyboard tray at the front. I have lowered the corner section of modular cubes as much as 5 in., so that the monitor is quite low, and installed a sit/stand keyboard tray on that section so that the tray will rise to match the previous height of the corner section and side surfaces. Although a gap is created with the corner section lowered and the corners of the adjacent side surfaces protrude into the gap, the keyboard tray moves the user back from those protruding corners so they pose no obstacle. The adjustments to the keyboard tray and chair will allow users to fine tune the optimal positioning for all aspects of their posture.

Employees Who Type With Excessive Force

Many employees will strike the keyboard keys so hard that an audible sound is heard from quite a distance. This unnecessary force can be damaging. Contact with the backspace key is often the most pronounced strike, even though they use digit 4 or 5 to press it. That an employee uses excessive force can often be confirmed by their closest neighbors (within hearing distance of the evaluation dialogue), who are frequently forthcoming on this aspect of close cube life. As usual, a lighthearted approach to the review will elicit more information than a strictly pragmatic methodology.

Reducing the striking force is not easy for users to do. In cases where the forceful typist was a pregnant woman with swelling in her wrists, I had success by recommending that her current keyboard be replaced by a "special keyboard" that absolutely required a soft striking touch or it would be damaged. The arrangement I make behind scenes was for another standard keyboard (which perhaps was identical to the keyboard she has) to be provided to her, and then reconfirming that this keyboard must be treated with a light touch. Such a deception may be the best way to change this habit, and it can be very successful when executed with confidence. Users end up reducing their force if they use the "special keyboard" for a week or more, and even when the ruse is exposed, they will continue to use a softer touch.

Employees With Only One Hand and/or With Limited Finger Dexterity

When a person types with only one hand, he or she necessarily incurs ulnar deviation when using the keys farthest from the active typing hand as well as the typing workload of two hands. Angling a standard keyboard to bring the distant side of the keyboard closer can be problematic because of limited desk or keyboard tray space, as it may not be deep enough for that positioning. Also, when the keyboard is angled in that fashion, a longer reach is created for the keys that are closest to the active hand.

There are an abundance of small keyboards available that work well in these situations, and many do not have an embedded numerical keypad, making them much less wide but with normal sizing otherwise. If a numerical keypad is needed, some small keyboards offer an external keypad as an accessory. Stand-alone numerical keypads are also common, and many have additional features such as ancillary function keys and/or secondary USB ports.

Also, since USB connectivity has proliferated, there are an abundance of specialized input devices that allow the programming of long strings of text into one key, or provide control of a mouse without using the hands. Also, voice recognition software has expanded its functionality and simplicity of operation, reducing hand and finger use significantly.

Employees With Restless Legs

I worked with a middle-aged woman who had some minor upper extremity discomfort but was really hoping to address restless leg syndrome, which had become exacerbated by some new medications she was taking. She explained that it now affected her during the late afternoons while working on her computer, in addition to the nighttime bouts she had endured for years. She described her discomfort as a sensation of tingling within her legs and a relentless compulsion to move them as she sat at her workstation. She sat in a cube that was part of an open group of five cubes, and she worried that her continual leg movements on and off the legs of her chair were a distraction to others, and she felt very self-conscious about it.

The solution was twofold. The first modification was to adjust her chair so that she received optimal leg support. The chair seat pan was extended out so that she had support all the way along her upper leg and to reduce any pressure points against her legs from the front edge of the seat pan. Second, she was provided with a massaging footrest. This is a fairly common style of footrest, where the center section of the resting platform is comprised of many rows of large massage beads that revolve on horizontal shafts. As the user supports their feet on the footrest, they can also roll their feet across the rotating beads to stimulate circulation, which is especially effective when not wearing shoes. Additionally, the model used had a "rocking chair" style of base, which allowed the platform to be rocked, forward and back, as the user's feet moved over the massage beads. Other models have a flat base that remains stable, but the platform rocks forward and back on the side rails that support it. Because the device was made to encourage and allow motion, this woman felt much more comfortable and less self-conscious about her leg movements.

EMPLOYEE SITUATIONS THAT ARE ARCANE BUT ERGONOMICALLY EDUCATIONAL

Window-Shade Neck

I often work with persons who have their monitors near windows, where a lot of light and glare affect the image on the screen. Frequently, these employees have problems from the proximity to the abundant natural light, but are willing to make concessions in order to have a pleasing view.

I encountered one employee with a stunning view of the front range of the Rockies (which is not uncommon) who was reporting significant neck discomfort, yet had a workstation that was almost ideal in the way that it was configured ergonomically.

As I worked with this person, overlooking the pristine eastern slope, I was at a loss for what could be causing his neck discomfort and had completed all of the typical evaluation tasks without identifying an apparent source. This man was quite tall and had even elevated his large CRT monitor to a perfect height, yet reported that his discomfort occurred only during work or right after the end of his workday. I asked this young man to continue working as I sat behind him, scratching my head as to what the cause might be. He was not prone to repeated or prolonged gazing out the window during his workday, which removed the prospect of tweaking his torso for a turn toward the windows, and I remained in the well-lit dark. As I sat, I tried my best to "take it all in" and noticed that just as a dark, moisture-laden cloud passed in front of the sun, his head and neck posture changed. I did not immediately make the connection, until I focused more on his screen and the spreadsheet he was scouring for information. Suddenly, I realized that the passing cloud had relieved his head from the function it served in relation to the profuse light through the windows—he used his head as a shade! The neck strain he was experiencing was due to continually positioning his head so that it shaded his specific viewing area on the large screen. As he viewed other areas of the screen, and as the sun changed positions in the sky, his "head shade" was repeatedly repositioned so that the glare was diminished in the exact area of the screen that he was reading. The result was a continual strain on his neck muscles to move his head and to maintain the precise position that blocked the light. As an immediate test scenario, I placed a large piece of cardboard from a shipping box in front of the glass, and he reported signs of less tension within a few minutes of physically realizing that the glare would not be a recurring phenomenon as he worked. The timing of the evaluation was accidentally instrumental, as the sunny weather had precipitated his discomfort, as it usually did, and the role it played as the cause was clearly displayed. He ultimately moved to a cube away from the windows.

Canine-Care Cramping

I was called to work with a woman who was a manager for a large company and had a wonderfully designed workstation but was having shoulder, arm, and wrist pain on her dominant side. She reported that the muscles in her upper arm would cramp up as she moved her right hand between her keyboard and mouse. As I conducted the initial information gathering, I lingered because I was aware that she had an adjustable keyboard tray set perfectly, a nice keyboard and mouse, and I began to wonder how her workstation could be causal.

The "ergo detective" phase kicked in and I began to scour the entirety of her desk and office for any clues. On the wall closest to her desk, I noticed the nicely framed individual photographs of three very different but very well-groomed dogs. I offered a compliment about these well-cared-for creatures and she lit up with pride. She became enthusiastic and began to demonstrate her daily routine of energetically brushing and grooming these dogs. She elevated her arms and extended her right (dominant) hand as if she was holding a brush and as she began to slide it across an imaginary afghan, she froze and her mouth dropped open. She and I simultaneously realized that it was not her mouse nor any component of her workstation that was the main cause of her discomfort, it was brushing her dogs. As it turned out, the photos were relatively new and she had prepared her dogs for the photo shoot about three weeks before requesting the evaluation. Of course, her computer work had also increased since she was doing employee reviews during the same period, but the computer work alone would not have triggered her discomfort so severely. This was a fine example of cumulative trauma being the cause.

SUV Shoulder

In the course of a two-year period, I worked with three women who were all 5 ft 2 in. or shorter, right-handed, and complained of severe left shoulder pain while demonstrating no egregious postures or motions in their work involving their left arm. My first encounter was dumbfounding, as the woman's workstation created wonderfully neutral postures, with minimal impact to her left arm and shoulder. I was at an impasse for the cause of her discomfort, and might not have uncovered the source if the employee had not made very frequent glances through her window toward the parking lot. I mentioned her recurrent glances as we talked and learned that her concern was for a dog inside her SUV. Since she was so diminutive, I invited a quick walk out to the car so that she could check on her pet, anticipating that the dog was very large and that she may have strained her left shoulder from some activity with it.

However, her approach to the large SUV clearly revealed the source of her left shoulder discomfort as she got in to greet the tiny dog sitting on the passenger seat. After opening the driver's door to enter, she reached her left arm up toward the left side of the windshield, where a vertical handle is installed on the left pillar post just inside the door. As she acquired the handle with her outstretched left hand, she raised her left foot up onto the running board and hoisted herself up by her left arm. Her weight, beyond the median for her height, was such that the left arm strain from this frequent maneuver was the cause of her discomfort and she unconsciously winced as she landed in the seat.

Because of the height of many large SUVs, short women have to lift themselves up and into the cab using the pillar post handles that some SUVs and pickup trucks install for that purpose. In the case I mentioned here, the strain of this action was often exacerbated by carrying a purse on her right shoulder and being a bit overweight.

Because SUVs are very common vehicles in Colorado, I later ran into exactly the same situation twice more after this. Two short women who presented with left (nondominant) shoulder discomfort (and also drove SUVs) were understandably taken aback when I posed the question, "Do you drive an SUV with a hand grip beside the windshield?" They were even more astonished when we investigated the significant impact that the left-hand handle played in their situations.

Bowling Shoulder

I received a call one day from an employee at the university who pleaded with me to come over right away because she was suddenly having horrible right shoulder pain. She described her pain as being extreme from her shoulder to her elbow and every time she moved her mouse made it worse! Because most ergonomics-related symptoms develop over time, I asked some questions about any changes in her recent work activities as I completed the subjective history section of the interview. She sat back and recounted her activities during the previous workweek and replied that "nothing had changed." Since her workstation was ergonomically sound, I paused so that I could gather my thoughts. I then asked about any recent outside activities. She replied that nothing new had transpired after work either. But suddenly she stopped for a moment, and then eagerly acknowledged that she had gone bowling on Saturday night for the first time in six years, bowling five games against a team of friends. She then looked down, laughed, and said that she now understood why her "mouse was causing such pain." But her team had won!

GLOSSARY

The terms listed here are in common use in ergonomics and should assist readers to understand their use in this volume.

Abduct: movement of an arm or leg away from the body.

Adduct: movement of an arm or a leg toward the body.

Anterior: referring to the front.

Cervical spine: the upper section of the spine; the neck. Seven vertebrae between the skull and the thoracic spine.

Contact pressure: pressure placed on a body part when it is being supported by a solid surface or when a solid surface is pressing against an external body area.

Dorsal: referring to the back or posterior.

Elbow extension: occurs when an arm is opened toward a straight position with minimal bend in the elbow.

Epicondylitis: the onset of pain on the outside (lateral) of the elbow; usually gradual with tenderness felt on or below the joint's bony prominence. Movements such as gripping, lifting, and carrying tend to be troublesome.

Ergonomist: title often associated only with practitioners certified by the Board of Certified Professional Ergonomists as a CPE. In this

Safety Manager's Guide to Office Ergonomics, By Craig Chasen
Copyright © 2009 John Wiley & Sons, Inc.

book the term is used to designate a person who conducts an ergonomic evaluation. At this time there is no licensing authority for those performing ergonomic evaluations.

Force sensors: small, inexpensive force-sensing resistors are small pads that can be connected to a simple voltmeter to measure the amount of force that a person uses when pressing on a solid surface. The typical applications are for measuring finger pressure on a keyboard, mouse, or calculator.

Ganglion cyst: a bump or mass that forms under the skin in the hand or wrist, most commonly on the back side. They can increase in size when the tissue is irritated and often "disappear" spontaneously. They may be rock hard due to the high pressure of the mucous-like fluid contained within the cyst and are often mistaken for a bony prominence.

Goniometer: an instrument that measures angles. The term is derived from two Greek words: *gonia,* meaning angle, and *metron,* meaning measure.

Hypermobility: occurs when a joint moves easily beyond the normal range expected for a particular joint. Also known as *joint hyperlaxity.*

Hypothenar pad: the fleshy mass at the medial side of the palm of the human hand beneath the little finger. Also called *antithenar.*

Kyphosis: rearward (outward) curvature of any part of the spine (usually thoracic).

Lateral: toward the outside; left or right of the body.

Lordosis: forward (inward) curvature of any part of the spine (usually lumbar).

Lumbar spine: the lowest segment of the spine consisting of five vertebrae (some individuals have six.)

Medial: toward the middle; nearer the midline of the body.

Mousing: the activity of using any type of mouse device.

MVA: motor vehicle accident.

Negative tilt: refers to a keyboard tray that is positioned with the back edge lower than the front edge.

Neutral posture: occurs when the body part is in its most relaxed position.

Popliteal height: the distance from the floor to the underside of the thigh when in a seated posture.

Posterior: referring to the back or behind.

Radial deviation: occurs when the hand is bent inward at the wrist, toward the thumb.

Radiculopathy: describes a condition in which one or more nerves are affected and do not work properly as the result of a problem at or near the root of a nerve along the spine. This can result in pain, weakness, numbness, or difficulty controlling specific muscles. Although caused near the spine, the pain or other symptoms may manifest in an extremity through the process of referred pain. For example, a nerve root impingement in the neck can produce pain and weakness in the forearm. It is also very common in the legs as the consequence of lower back problems.

Recovery time: a period of time wherein muscles and tendons that were under strain or exertion can be relieved of engagement for rest and replenishment.

R > L, L > R: right greater than left or left greater than right: to indicate discomfort that affects both sides, as in arms or wrists, but one side is greater than the other.

Scapula: the shoulder blade.

Sciatica: a set of symptoms referring to a burning, stinging, and/or numbing pain that is felt in the buttock, thigh, leg, and/or foot, which may or may not be associated with lower back pain. Typically, the symptoms are felt on only one side of the body.

Scoliosis: a condition where the spine is curved side-to-side or rotated. The spine, from the back, looks more like an "S" than a straight line.

Shoulder abduction: moving the arm away (outward) from the torso laterally.

Shoulder adduction: moving the arm across (in front of) the torso laterally.

Shoulder flexion: shoulder is elevated above a neutral posture.

Solutions: equipment recommended to correct an ergonomic problem area.

Static posture: a body position fixed over time, with muscle contraction but without motion.

Thenar pad: (or thenar eminence or *prominence*): the body of muscle in the palm of the human hand beneath the thumb.

Thoracic spine: the middle segment of the spine with 12 vertebrae between the lumbar spine (lower back) and cervical spine; neck.

Ulnar deviation: occurs when the hand is bent outward at the wrist, toward the little finger.

Volar: referring to the palm of the hand or the sole of the foot.

Wrist extension: occurs when the hand bent back (upward) at the wrist.

Wrist flexion: occurs when the hand bent forward (downward) at the wrist.

INDEX

Printed in the United States
By Bookmasters